FUNDAMENTALS OF COMPILERS

AN INTRODUCTION TO COMPUTER LANGUAGE TRANSLATION

KAREN A. LEMONE

CRC Press
Boca Raton Ann Arbor London Tokyo

Acquiring Editor: Russ Hall
Production Director: Sandy Pearlman
Production Services: HighText Design and Production
Cover Design: Shi Young

Library of Congress Cataloging-in-Publication Data
Lemone, Karen A.
    Fundamentals of compilers : an introduction to computer language
translation / Karen Lemone.
        p.   cm.
    Includes bibliographical references and index.
    ISBN 0-8493-7341-7
        1. Compilers (Computer programs).   I. Title.
P308.L45   1992   91-26325
418′.02′0285—dc20   CIP

Direct all inquiries to CRC Press, 2000 Corporate Blvd., N.W., Boca Raton, Florida, 33431.

©1992 by CRC Press, Inc.

International Standard Book Number 0-8493-7341-7
Library of Congress Card Number 91-26325
Printed in the United States of America   2 3 4 5 6 7 8 9 0

**Dedicated to my daughter Jamie Lemone
who teaches me things all the time**

ABOUT THE AUTHOR

Dr. Karen A. Lemone is Associate Professor of Computer Science at Worcester Polytechnic Institute, Worcester, Massachusetts. She has been teaching compiler construction courses for over ten years and has served as interim department head of the Computer Science Department at Worcester Polytechnic Institute. She is currently a site-visitor for the Computer Science Accreditation Board (CSAB).

Dr. Lemone received her B.A. degree in Mathematics from Tufts University, M.A. degree in Mathematics from Boston College, and Ph.D. degree in Mathematics/ Computer Science from Northeastern University.

In the summer of 1985, she was invited as a foreign expert to Jilin University of Technology, Changchun, China. For the academic year 1988–1989, she was a *professor invité* at Ecole Polytechnique Fédèrale de Lausanne in Switzerland.

Dr. Lemone has published many articles in computing publications and has co-authored several books in computer science. She has been a reviewer for ACM *Computing Reviews* and IEEE *Computer,* as well as for conferences and various book publishers.

Her current research interests include semantic analysis and optimization techniques for object-oriented languages, applications of attribute grammars, and electronic document processing. She consults and conducts seminars in compiler implementation.

CONTENTS

3. Introduction to Grammars and Parsing

4. Top-Down Parsing

5. Bottom-Up Parsing

6. Error Handling

7. Semantic Analysis

8. Symbol Tables

9. Introduction to Code Generation

PREFACE

The design and implementation of a *compiler* is an example of a large, complex programming project. Good compiler design and implementation requires the best knowledge of software engineering in order to handle the complexity.

The design and implementation of a compiler *text* is an example of a large, complex writing project. Good textbook design and implementation requires the best knowledge of educational pedagogy in order to handle the complexity.

There are many well-designed and well-implemented compilers today, which translate efficiently and are well modularized to deal effectively with the complexity of language translation.

These two texts, *Fundamentals of Compilers: An Introduction to Computer Language Translation* (referred to here as *Book I*) and *Design of Compilers: Techniques of Computer Language Translation* (referred to here as *Book II*), are modularized to deal with the pedagogical complexity of compiling. Principles are stressed whenever possible, with many, many examples to support and clarify them.

Book I discusses the standard phases of a compiler: lexical analysis, syntax analysis, semantic analysis, and elementary code generation. Optimization is touched upon, but not treated in depth. Book I includes a project to create a compiler for a subset of Ada. No knowledge of Ada is required.

Book II discusses the standard phases of a compiler, but in greater depth than in Book I. Lexical and syntax analysis are discussed from a compiler generator point of view. Chapter 1 reviews the basic concepts and describes algorithms for *generating* scanners and parsers. Other topics in Book II include attribute grammars and their use in semantic analysis, algorithms and data structures for optimization and code generation, object-oriented languages, production-quality compiler issues, and compiling for special architectures.

Both Book I and Book II contain a mixture of practical and (appropriate) theoretical information. I think students should be exposed to both. Issues that arise from language, grammar, and theory are discussed in relationship to the compiling process as they occur, rather than in separate chapters. Discussing these issues in their proper context allows students to see their relevance.

These books conform to the ACM and IEEE Curriculum recommendations for compiler courses.

Possible course uses include:

- A full semester course (*easiest*): Book I: Simple exercises, omitting perhaps Chapter 5. Project: A recursive descent compiler (described in the addendum to each chapter) omitting Project, Part V (the error handling routines of Chapter 6). Include Project, Part IV even if Chapter 5 is omitted.

- A full semester course (*moderate difficulty*): Book I: All chapters, including some of the more difficult exercises, either assigned or discussed in class. Chapters 1 and 2 from Book II, more as time and interest allow. Project: Either a recursive descent compiler (perhaps including Part V, the error handling routines of Chapter 6) or one using a compiler tool such as LEX/YACC.

- A full semester course (*intensive*): Book I and Chapters 1 to 10 of Book II. Project: Either of the compiler options described in *easiest* and *moderate difficulty,* or see the project outlined below for two-term sequence courses.

- Two shorter 7- to 10-week terms:

    Term I: Book I, Chapters 1 to 8, omitting Chapter 5 for a somewhat easier course. Project: Parts I–VII, omitting Part V, if desired. Do not omit Part IV even if Chapter 5 is omitted.

    Term II: Book II, Chapters 1 to 10. Project: Reimplement the compiler project (1) *using* a compiler generator or (2) *writing* a top-down parser generator as described in the addendum to Book II, Chapter 1; continue with the projects as assigned at the end of each chapter in Book II.

- A graduate course: Book II, supplemented with the Related Reading suggestions included at the end of each chapter. Project I if the students have no previous compiler background; Project II if they have some compiler background. Both projects should use compiler tools such as LEX/YACC or create their own (top-down) parser generator.

ACKNOWLEDGMENTS

Many people and many classes have contributed to the development of this text. I would like to especially acknowledge the following people: John Casey, Larry Engholm, Enya, Patty Gigliotti, Gary Gray, Gary Gu, Marty Kaliski, Ellen Keohane, Kitaro, Patti Lynch, Alok Mishra, Mat Myszewski, Kathy O'Donnell, David Paist, Diane Ramsey, Michael Smith, Jeff Smythe, Mib Stancl, Del and Sue Webster, Jim Whitehead, CS 544, Fall 1990, CS4533 C Term, 1991.

FUNDAMENTALS OF COMPILERS

AN INTRODUCTION TO COMPUTER LANGUAGE TRANSLATION

1

Overview of Compiling

1.0 Introduction

There is an interesting dilemma in the computer world: A gap exists between the kind of thinking that people use to solve problems and the way computers are designed to solve problems.

Computers are machines with a memory, designed to operate on data by shifting bits, performing arithmetic, logical and sometimes character operations, with various constructs for looping and controlling what operations are to be performed next in a sequence of operations.

Research into problem solving and into the ways humans think has resulted in programming languages which allow us to abstract our thinking and write problem solutions top-down, using methods of step-wise refinement. Newer object-oriented approaches continue this abstraction process. Much has changed in this domain.

Research into computer design has resulted in new, more efficient and more complex architectures to operate on bits, performing arithmetic, logical and sometimes character operations, with various constructs for looping and controlling what operations are to be performed next in a sequence of operations—that is, not much has changed in this domain.

Thus, as languages evolve to try to capture human problem solving ability, computers remain, in this respect, very much the same. In addition, the newer architectures, such as those with reduced instruction sets and those for parallel processing, are more complex to program. Steadily, the gap widens.

Language translators bridge this gap.

1.1 Compilers

A *compiler* is a special translator which takes a high-level language, such as Pascal or C (or your favorite high-level language), and translates it to a low-level representation which the computer can ultimately execute. In this text, we will consider compilers which translate from high-level language to assembly language. The assembler can then finish the translation process and the system linker/loader can prepare the program for execution.

High-Level Language → Compiler → Low-Level Language
Program   Program
(e.g., Pascal)   (e.g., assembler)

Figure 1

1.2   History of Compilers

The word *compiler* is often attributed to Grace Murray Hopper, who visualized the implementation of a high-level language, quite accurately at the time, as "a compilation of a sequence of subroutines from a library." The first efforts at implementing English-like languages (or French-like for those who speak French) were done in this way.

The first translator-like compilers were written in the late 1950's. Credit is often given to FORTRAN as the first successfully compiled language. It took 18 person-years to develop the FORTRAN compiler because the language was being designed at the same time that it was being implemented, and since this was a first all around, the translation process was not well understood.

John Backus headed the design team which invented and implemented FORTRAN. It is hard for us today to appreciate the significant step represented by FORTRAN and the FORTRAN compiler.

At the time, there was great skepticism that an English-like language could be designed and translated into anything which the machine could execute efficiently. The fact that problems could be described better in high-level languages using sophisticated data structures and control constructs was only beginning to be grasped. The saying

> Real programmers use assembly language

was firmly believed (although not expressed this way!) by many computer people at the time. To be truthful, a good assembly language programmer can, even today, write better code in some cases than a compiler can produce, but the advantages of solving a problem using the abstractions found in high-level languages far outweigh this in all but a minority of cases.

The FORTRAN compiler was designed to put out the very best code possible (that is, code which would execute fast) to overcome the skepticism that an English-like language would be too inefficient. FORTRAN was designed with the need to prove itself, so that many of the features in the language reflect the design for efficiency. FORTRAN lacks the sophisticated data structures (e.g., records, sets, pointers, enumerated types) of newer languages such as Pascal and Ada; in addition, recursion is not built into the language FORTRAN (it must be simulated using a stack).

We can now design and implement a compiler faster and better for many reasons. First, languages are better understood. The FORTRAN language was designed at the same time as its compiler. Second, tools have been developed to ease some of the tasks performed—particularly by the first phases of a compiler. Third, data structures and algorithms have been developed through the years to perform the tasks common to all compilers. Finally, experience with languages and compilers has resulted in the design of languages and language features for which it is easier to write compilers.

1.3   Compiler Phases

The compiling process can be described as a sequence of serial phases starting with lexical analysis and ending in code generation, with table handling, error and I/O modules interacting with more than one phase:

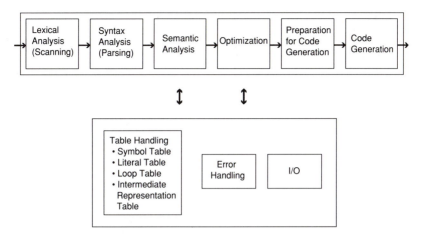

Figure 2

Most of the terms in this picture will be discussed in this chapter. This diagram is a simplification of the compiling process since many of the phases interact with each other or are not always performed in the order shown. For example, many compilers are "parser-driven" (*parsing* is another name for the syntax analysis phase) which means that the syntax analyzer (parser) calls the lexical analyzer, rather than the lexical analyzer completing its task before the syntax analyzer begins.

The real picture is more like this:

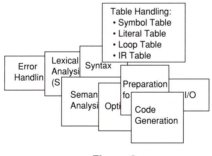

Figure 3

Clearly, this picture, although perhaps more representative of real compilers, is not conducive to an initial understanding of compilers. The original, serial picture, is much easier to understand and even though it may be a less efficient implementation, it is still accurate. A compiler designed this way would still execute correctly albeit more slowly than a so-called "production quality" compiler.

In the remainder of this chapter, we will take a small two-line Pascal-like "program" through the various phases, showing what is produced along the way. The discussion here is an overview of *what* a compiler does. The remainder of this book will discuss *how* these tasks are accomplished.

EXAMPLE 1   A simple compiler problem

Consider the following program consisting of two assignment statements as input to a compiler:

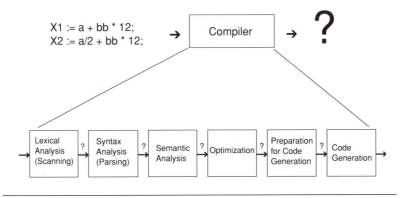

We will investigate what the outputs are from the various phases; that is, we will determine what is represented by the question marks.

The first phase to be considered is lexical analysis.

1.4   Lexical Analysis

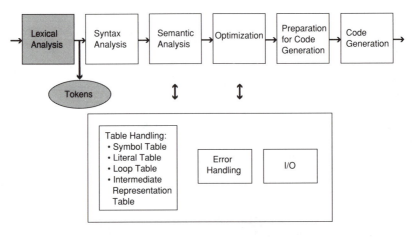

Figure 4

Lexical analysis, also called *scanning* or sometimes *scanning and screening* or, even less formally, *lexing* or *tokenizing*, groups sequences of characters from the input source program into units called *tokens*. Thus, lexing can be considered as a mapping:

Sequence of Characters → Sequence of Tokens

1.4.1   Tokens

Tokens are the basic lexical units much as words and punctuation are the basic language units of an English sentence. Tokens vary from language to language and even from compiler to compiler for the *same* language. Choosing the tokens for a language is one of the tasks of the compiler designer.

For our purposes we will consider these basic lexical units or tokens to be special sequences of characters which comprise the "words" of our language. A token is the smallest language unit which conveys meaning.

In lexical analysis, the sequence of characters is *scanned* until a legal sequence is found; it is then *screened* to classify it by a *type* such as *integer* or *keyword* or *identifier*, etc.

Some examples of tokens are:

- A keyword *IF*
- A constant *12*
- An identifier *X1*
- An operator *<* or *.LT.*
- Punctuation *(* or *;*

These examples are language dependent. Languages such as FORTRAN do not use semicolons, so there would be no valid token for a semicolon. On the other hand, many lexical analyzers for languages that contain both keywords and identifiers, such as Pascal or C or Ada, might classify both as identifiers. A later phase of the compiler would then separate them into their specific types or classes.

In

```
IF New > MaxNum THEN •••
```

the tokens are:

```
"IF"
"New"
">"
"MaxNum"
"THEN"
```

Tokens generally are described in two parts, a *type* or *class* and a *value:*

Token = (Type, Value)

For the sequence above, the tokens might be described as:

```
(Keyword, IF)
(Ident, "New")
(Oper, >)
(Ident,"MaxNum")
(Keyword, THEN)
```

where "IF" and "THEN" have been classified as keywords, "New" and "MaxNum" as identifiers, and ">" as an operator.

1.4.2 Other Tasks Performed by the Lexical Analyzer

Lexical analyzers perform other tasks such as deletion of blanks and comments; they sometimes build a symbol table (or name table—a table of the user-defined names used in a program). In this book, we will consider the task of building the symbol table as part of the later, semantic analysis phase. Simple errors such as misspellings can also be found by the lexical analysis phase.

We can now look at our two assignment statement program and see the results after lexical analysis.

EXAMPLE 2   The simple compiler problem—lexical analysis

Here is our two-line example before and after lexical analysis.

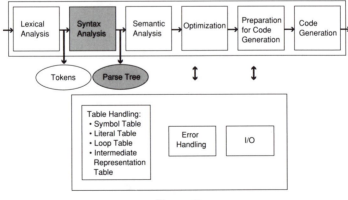

Here, *X1* has been classified as an *Id* whose value is the string "*X1*"; the assignment operator, :=, has been classified as an *Op*; "*a*" is another *Id*; 12 and 2 are classified as *Lits* (*literals*), etc. When we look at this example, we can see why the output of the lexical analyzer is often described as a "stream" of tokens.

1.5 Syntax Analysis or Parsing

Figure 5

Syntax analysis is more complex than lexical analysis. The syntax analysis or parse phase of a compiler groups tokens into syntactic structures much as we had to structure sentences back in eighth grade. In sentence structuring,

The little boy ran quickly

has a noun phrase (*The little boy*) and a verb phrase (*ran quickly*). The verb phrase consists of substructures, a verb (*ran*) and an adverb (*quickly*). The words are analogous to the tokens found by the lexical analyzer. The noun phrase, the verb phrase, the verb, the adjective and the adverb are analogous to the structures found by a syntax analyzer.

For a programming language,

```
bb * 12
```

consists of three tokens (*bb*, *, and *12*) and might be grouped into a structure called an *expression*, while

```
X1 := a + bb * 12;
```

might be grouped into a structure called an *assignment statement* which contains the substructure *expression*.

In the process of finding the syntactic structure, the syntax phase also determines if a sequence of tokens is syntactically correct, according to the definition of the language.

Just as tokens may consist of one or more characters, syntactic categories may consist of one or more tokens.

The structure recognized by syntax analysis is described in a tree-like form called a *syntax tree*, *parse tree* or *structure tree*. Not all parsers produce a tree representation explicitly, but the output can always be used to create a parse tree.

Following is a syntax (parse) tree after syntax analysis of our two-ine example. The structure is similar to the structure in arithmetic: a *term* is a product and each part of this product is called a *factor*. A sum of terms and factors is called an *expression*.

EXAMPLE 3   Our two-line example before and after parsing

```
X1 := a + bb * 12;
X2 := a/2 + bb * 12; → (lexical analysis)
```

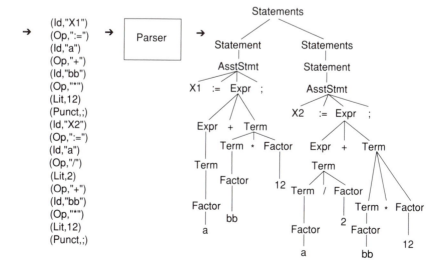

Creation of this structure from the token stream is the subject of Chapters 4 and 5. Notice that the leaves of the tree are tokens; non-leaves are syntactic categories. For clarity, we have shown only the input characters on the leaves rather than the ordered pair consisting of the type and value. For Example 3, the left-most leaf, *X1*, is a token such as *(Id, "X1")*.

In the above tree, a program consists of statements; each statement is a single statement followed (optionally) by more statements; the statements are identified as assignment statements (**AsstStmt**). The first assignment statement consists of four parts: (1) the token representing **X1**, (2) the token representing the assignment operator, **:=**, (3) the syntactic category expression, **Expr**, and (4) the token representing the semicolon, **;**. In turn the **Expr** here is grouped into three parts: (1) the syntactic category **Expr** again, (2) the token representing **+**, and (3) the syntactic category **Term.**

Clearly, there are a lot of syntactic categories for such simple assignment statements. These categories are part of the language's grammar. At the back of most programming language books is the grammar for the language. Pascal's grammar is often described using syntax diagrams. Grammars are introduced in Chapter 3.

There are many algorithms, both simple and complex, for creating parse trees. Not too surprisingly, the simple ones tend to be less efficient than those which are more complex. We will study a number of algorithms for creating parse trees in Chapters 3, 4 and 5.

1.6   Semantic Analysis

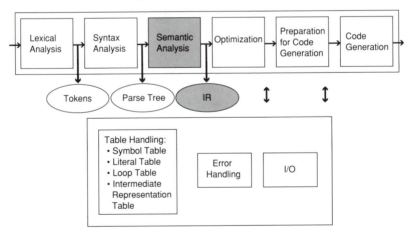

Figure 6

The semantic analysis phase follows the parse phase and takes as its input the parse tree created in the syntax analysis phase.

This phase determines the meaning (semantics) of the program by creating symbol tables, checking that variables used have been defined (languages like Pascal require this), and a myriad of other tasks prior to generation of code.

Two of the main tasks in the semantic analysis phase are (1) *static checking* of the program and (2) generation of an *intermediate representation (IR)*.

1.6.1   Static Checking

Static checking completes the analysis begun by the parser and performs activities such as affirming that a character-valued variable isn't being assigned to a variable declared to be an integer-valued variable. This is called *type checking*.

EXAMPLE 4   Static type checking

Is the expression *a* ∗ *b* + *c* legal?

Although we might initially say *yes*, the answer would be *no* in most languages if *a*, *b*, and *c* are declared as logicals (Booleans)! Thus, the semantic analyzer will check that ∗ and + are legal operators for the variables *a*, *b*, and *c* by checking their declared type.

Another static check is an array dimension check: An array declared to be two-dimensional can't be used with three indices, for example. Still another static check would confirm that all variables used have been declared if the language requires this (not all do).

Static type checking varies with the language being compiled. An error in one language might not be an error in another.

1.6.2   Intermediate Representation (IR)

Intermediate representation, sometimes called Intermediate Language (IL) or Intermediate Code (IC), is an alternative form to a parse tree. Sometimes the parser creates this intermediate representation directly; sometimes the parse tree is converted to the representation.

Here, we show a form called an *abstract syntax tree* or *abstract structure tree*. It removes some of the intermediate categories and captures the elementary structure. Each leaf *and* non-leaf represents a token; leaves are operands and non-leaves are operators. The assignment statement:

```
X := X + 1
```

has an abstract syntax tree:

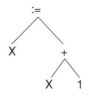

Figure 7

An abstract syntax tree for:

```
     IF (A < B) THEN X := X + 1
```
is:

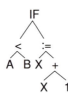

Figure 8

The exercises at the end of the chapter (see Exercise 8) discuss abstract syntax trees for other common programming language statements.

EXAMPLE 5   Our two-line example converted to an abstract syntax tree

```
X1 := a + bb * 12;
X2 := a/2 + bb * 12;  → (lexical analysis, syntax analysis)
```

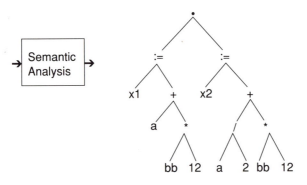

Abstract Syntax Tree

For simplicity, the input is shown as the two original statements rather than the parse tree produced by the syntax analysis phase.

In Chapter 6, we will see other intermediate forms used by compilers.

1.7   Optimization

The optimization phase changes the intermediate representation so that the later code generation phase will produce code which will execute faster or take up less space (or both).

1.7.1   Types of Optimizations

Four types of optimizations can be identified: (1) local optimizations which are made within a statement or group of statements, (2) loop optimizations which are performed inside loops, (3) global optimizations which are performed over an entire program or procedure and (4) "peep-hole" optimizations—those performed *after* code is selected by "peeping" at a small sequence of code.

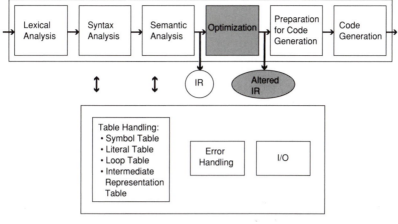

Figure 9

We will show two local optimizations here. The first is called constant propagation and the second common subexpression elimination.

Constant Propagation

Consider the following two statements:

```
X := 3
 .
 .
 .
A := B + X
```

Of course, the optimization phase actually operates on an intermediate representation:

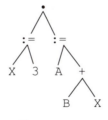

Figure 10

but we will look at the program for simplicity.

The code generation phase creates the assembly language, and assembly language code for this might look something like:

```
Move #3,X ; Move the constant 3 to X
 .
 .
 .
Add X,B,A ; Add X to B and store it in A
```

(The reader can translate this into his or her favorite assembly language.)

Each reference to a variable name such as *X* requires the computer to execute a "fetch" from memory. When the 3 is contained within the instruction itself, the computer can often compute the value faster.

Suppose it is true that the program always executes the second assignment statement after it has executed the first; that is, there are no branch statements that ever cause the program to execute the first and not the second or the second and not the first. Then, we can change the program to:

```
X := 3
    .
    .
    .
A := B + 3
```

(Again, the change is really to the intermediate representation.)

Then, the assembly language code for executing this would look something like:

```
Move #3,X ; Move the constant 3 to X
    .
    .
    .
Add #3,B,A ; Add 3 to B and store it in A
```

Here, the constant 3 is substituted for *X*. This will execute faster because one less memory fetch is needed. This may not seem important, but if this sequence of code were inside a loop that executed many times, considerable execution time could be saved.

Common Subexpression Elimination

Consider the following statements:

```
A := B * C
    .
    .
    .
D := B * C
```

Let us assume that an algorithm has determined that the second statement is always executed if the first is and that no change has been made to *B* and *C* in between the two statements.

A common subexpression elimination algorithm might change this (again, the change would be to the intermediate representation) to:

```
T := B * C
A := T
    .
    .
    .
D := T
```

Although the resulting code clearly takes up more space, it will execute faster because a computer can copy information faster than it can perform a multiplication.

EXAMPLE 6   Optimizing our two-line example

Optimization operates on whatever intermediate representation is produced by the semantic analysis phase. Since the abstract syntax tree is difficult to read, we will show the intermediate representation as a sequence of assignment statements with one operator. (It is easy to convert an abstract syntax tree to this form.)

```
X1 := a + bb * 12 ;
X2 := a/2 + bb * 12 ;   → (lexical analysis, syntax analysis,
                              semantic analysis)
```

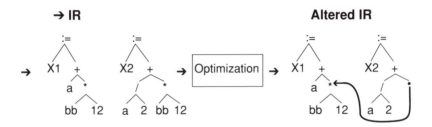

Here, the optimizer has performed common subexpression elimination; the resulting code will have only one instruction multiplying the quantity *bb* * *12*. Notice that the altered IR is a graph rather than a tree.

1.8   Preparation for Code Generation

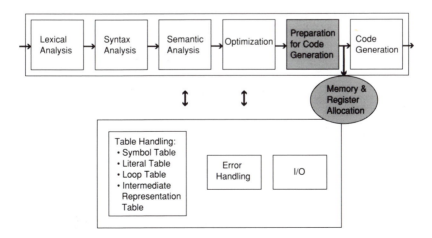

Figure 11

When a program is executed, the values for its variables and expressions are stored in memory and registers. The compiler decides many of the issues about where these values will reside. Two issues in preparing for code generation are, thus, *memory allocation* and *register allocation*.

Register/memory allocation and selection of assembly language code are a chicken and egg problem. If we select registers first, this will influence the resulting code that we select. If we select code first, this will often influence how we are able to use registers. Some compilers deal with this issue by selecting temporary instructions, then assigning registers and memory and finally returning to select better code. Here, we will assign registers and memory and then proceed to code generation.

1.8.1   Memory Allocation

Assigning and maintaining space in memory for holding the values of variables and expressions is called memory allocation.

One issue in memory allocation is whether memory is to be allocated statically, that is, fixed at compile time, or whether storage can be on a stack or heap whose size changes as the program executes.

1.8.2   Register Allocation

Registers are used to hold the values of variables and expressions. There are rarely enough for an entire program although the newer RISC (Reduced Instruction Set Computer) architectures have more. Programs execute faster when operations are performed on data in registers rather than on data in memory. Thus, a compiler would like to keep the values that are accessed the most in registers. Determining which values to keep in registers is difficult.

Good register allocation is often more important for the efficiency of the executable code than good optimization. Example 7 shows this phase for our two-line example.

EXAMPLE 7   Preparing our example for code generation

```
X1 := a + bb * 12 ;
X2 := a/2 + bb * 12 ;   → (lexical analysis, syntax analysis,
                                semantic analysis, optimization)
```

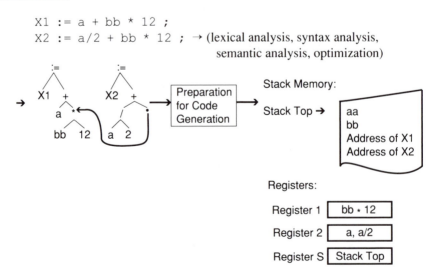

Here, the expression *bb* * *12* has been assigned to register 1 and a copy of *a*'s value is assigned to register 2. The value of *a/2* is also assigned to register 2. The variables are also allocated on the stack, with *a*'s value on top, *bb*'s value next, *X1*'s address next and *X2*'s address on the bottom. Register S points to (contains the address of) the top of the stack.

1.9 Code Generation

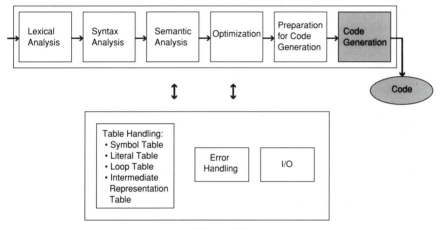

Figure 12

In this text, code generation means the translation of the intermediate representation to assembly language after optimization and after registers and memory have been allocated.

The main issue in code generation is the selection of "cunning" instruction sequences. Although mysterious but clever sequences of code should be well documented, putting out "readable" code is not the primary goal of code generation. The resulting code should execute correctly and efficiently. Example 8 generates code for our two-line example.

EXAMPLE 8 Code generation for our example

```
X1 := a + bb * 12 ;
X2 := a/2 + bb * 12 ;  → (other phases)
```

```
                          PushAddr    X2          ;Put address of X2 on stack
                          PushAddr    X1          ;Put address of X1 on stack
      ┌─────────────┐     Push        bb          ;Put bb on stack
 ───▶ │    Code     │ ──▶ Push        a           ;Put a on stack
      │ Generation  │     Load        1(S),R1     ;bb → R1
      └─────────────┘     Mpy         #12,R1      ;bb * 12 → R1
                          Load        (S),R2      ;a → R2
                          Store       R2,R3       ;Copy a → R3
                          Add         R1,R3       ;a + bb * 12 → R3
                          Store       R3,@2(S)    ;X1 ← a + bb * 12
                          Div         #2,R2       ;a/2 → R2
                          Add         R1,R2       ;a/2 + bb * 12 → R2
                          Store       R2,@3(S)    ;X2 ← a/2 + bb * 12
```

The assembly language code here is a pseudo-code; that is, it doesn't exist (to the author's knowledge) on any machine. The instruction mnemonics should be reasonably clear: **PushAddr** for push address and **MPY** for multiply, etc.

Because S contains the address at the top of the stack, we have used an indirection operator ( ) to get the actual contents pointed to by S. Thus (S) accesses the value of *a* from the top of the stack, and 1(S) accesses the value of *bb* one byte down in the stack.

The notations 2(S) and 3(S) reference the values at the third and fourth positions on the stack (since (S) refers to the value at the top). These positions contain addresses. To store at these addresses rather than on the stack, another indirection operator, @, is added. Thus "Store R3,@2(S)" means store the contents of register 3 in the location pointed to by S + 1. This is location *X1*.

The assembly language is not important here. What is important is that the compiler can produce executable code from the intermediate code and the storage and register allocation.

1.10   Summary

This chapter has described the various compiler phases and some of their functions.

The compiler is divided here into six phases, the first three of which are often termed the front end or analysis phases and the last three the back end or synthesis phases. The front end of a compiler is somewhat language dependent and the back end is more machine dependent. Thus, language issues are also discussed in the early chapters and machine issues are discussed in the later chapters.

The first phase, lexical analysis, finds tokens (the "words") in a program. The second phase, syntax analysis, finds the structure of a program as described by its grammar. The third phase, semantic analysis, finishes analyzing the program and translates it to an intermediate form for the synthesis phases. The fourth phase, optimization, finds ways to reduce the time or space to be used when the translated program is executed. The fifth phase, preparation for code generation, assigns memory and registers to hold the values and addresses of variables and expressions at run time. The sixth phase, code generation, produces assembly language code from the (optimized) intermediate representation.

The remainder of the book devotes a chapter or more to the algorithms and data structures which accomplish these functions for each phase. Moreover, there are more details concerning each function.

EXERCISES

The following questions are to be answered using only the information in this chapter. Some of these questions will be repeated in later chapters when answers can be more precise.

1. Match each of the following compiler functions with the phase that performs it.
   (a) Assigns a variable to register 5
   (b) Identifies *loop* as a label
   (c) Changes $A + 4 * 3$ to $A + 12$
   (d) Finds a variable that has not been declared
   (e) Changes $A := A + 12$ to *Add #12, A*
   (f) Creates a parse tree

   (i) Lexical Analysis
   (ii) Syntax Analysis
   (iii) Semantic Analysis
   (iv) Optimization
   (v) Preparation for Code Generation
   (vi) Code Generation

2. Identify some tokens in your favorite computer language. Create your own names for the types. For example, if your language has arithmetic operators, say, +, *, -, /, you might have a type called *ArithOp* with these four values or you might identify two types *MulOp* with values * and / and *AddOp* with values + and −.

3. Following the example in this chapter, create
   (a) a parse tree
   (b) an abstract syntax tree
   for the assignment statement *Max := 2 * Min ;*

4. (a) Translate the intermediate code from Example 6 before optimization to your favorite assembly language.
   (b) Translate the intermediate code from Example 6 after optimization to your favorite assembly language.

5. Assume the compiler for our two-line example is executing on a RISC machine that has 32 registers. What might the resulting code look like?

6. Show the results after each of the six phases for a program consisting of the single assignment statement: *Max := Min + 4 * 3 ;*

7. What is the assignment statement represented by the following abstract syntax tree?

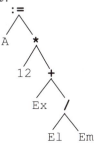

8. Following the model of the IF Statement whose abstract syntax tree is shown in Section 1.6.2, create an abstract syntax tree for:
   (a) If A < B THEN Min := A Else Min := B
   (b) While Count < 100 DO Count := Count + 1

9. For each of the following, give another term or word with the same meaning:
   (a) Token
   (b) Syntax analysis
   (c) Parse tree
   (d) Intermediate representation
   (e) Abstract syntax tree
   (f) Analysis phases
   (g) Synthesis phases

10. Bootstrapping: Suppose we want to create a compiler for (a) a new language on a machine that has compilers for other languages or (b) a new machine for a language for which a compiler exists on another machine. Consider each case separately:
    (a) Language L is new to machine M. Language L has no compiler. There is, however, a compiler for language Y on machine M. One can create a compiler for L in two steps:

        Step 1: Using language Y, write a compiler that translates (compiles) a small subset L_0 of language L to the assembly language of machine M.

        Step 2: Using the subset L_0, write a compiler that translates all of L to the machine language of M.

        Show how to use these programs to create a compiler that is written in the assembly language of M and compiles L programs to the assembly language of M.

    (b) Machine N is brand new. There are no compilers for any languages. However, there is a compiler for language L (written in L_0, a subset of L) on machine M. A compiler for L can be created for machine N in two steps:

        Step 1: Change the code generator of the L compiler on M so that it puts out code for machine N rather than M.

        Step 2: Run the whole new compiler through the old compiler. That is, take the source for the compiler with the new code generator and compile it using the L compiler as a program on M putting out M assembly language.Then we have a (cross) compiler on M, written in M that converts L programs to N's assembly language.

        Show how to use these two steps to create a compiler for L written in the assembly language of N that translates to the assembly language of N, i.e., a compiler for L on machine N.

2

Lexical Analysis

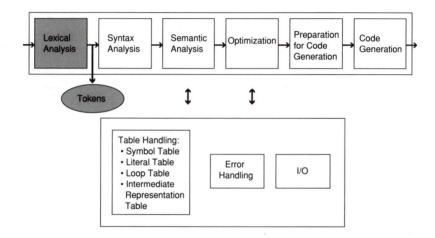

2.0 Introduction

The lexical analysis phase of a compiler groups sequences of characters into categories. Although this is also called *scanning*, we can actually distinguish two separate tasks: *scanning* and *screening*.

A *scanner* moves a pointer through the input one character at a time to find contiguous strings of characters which constitute individual textual elements (words) and to classify each according to its type.

The *screener* discards some of the tokens found by the scanner (perhaps spaces and comments), determines that others are reserved symbols (perhaps keywords and operators), and puts text of others in the name table.

The lexical analyzer passes to the syntax analyzer the type of the token, plus the actual value of the token. Sometimes, the value is a pointer to a table which contains the value. There can be separate tables for names, numeric constants, character string constants, and operators, or there can be one table including all of these.

2.1   The Lexical Analysis Problem

We can describe the lexical analysis problem as:

> *Given a string of characters, divide it into a string of tokens:*
>
> *Token = (type, value)*

The type of a token is often called its *class*; the actual characters are called the *lexeme*. There are three common formalisms or models for describing token types:

1) *Regular expressions.* Regular expressions describe tokens as the set of legal strings in a language and are described in Section 2.8.

2) *Transition diagrams.* Transition diagrams describe tokens as the legal strings which take the diagram from an initial state to a final state and are described in Section 2.5.

3) *Right linear grammars.* Right linear grammars describe tokens as the strings generated by a grammar in a special form (also called a *regular grammar*). Grammars are described in Chapter 3, but right linear grammars are a special form not described in this text (see Davis and Weyuker, 1983).

Technically, lexical analysis is not needed; it could all be done by a syntax processor (the next phase). There are two reasons for having a separate lexical analyzer, however. First, it is more efficient to have a separate lexical analyzer. Reading and analyzing input characters is a slow process, and it is important to make this phase as efficient as possible. Using a syntax analyzer is too general—it is like shooting sparrows with a cannon! Second, the resulting compiler is more modular.

2.2   Examples of Tokens

Our first example describes tokens for identifiers. Consider:

```
MaxNum = (Identifier, → MaxNum)
```

Here, *MaxNum* is identified as a token whose type is *Identifier* and whose value is a pointer to a table where *MaxNum* is listed as one of the user-defined names in the program. If this were position 72 in the table, we might say:

```
MaxNum = (Identifier, 72)
```

Many languages have special identifiers known as keywords. For example, the token

```
IF = (Keyword, "IF")
```

expresses the fact that we have a token whose type is keyword and whose value is "IF". If we had prestored a list of keywords in a table, we might return a pointer to this entry as the token's value:

```
IF = (Keyword, → IF)
```

or if it were entry 13:

```
IF = (Keyword, 13)
```

Again, sometimes keywords will be classified as identifiers during the lexical analysis phase and then separated out as keywords in a later phase:

```
IF = (Identifier, "IF")
```

Still another way to categorize keywords is to put each keyword in a class by itself; in this case we could have a type, say IF, with no value:

```
IF = (KeywordIF, __ )
```

Thus, tokens depend on both the language and the compiler designer's implementation.

2.3   Recognizing Tokens

Recognizing tokens is difficult in some languages. For example, in FORTRAN, the following represent legal statements:

```
DO 15 I = 1,25
DO 15 I = 1.25
```

The first statement is a looping construct called a DO loop. The second is an assignment statement. (The value *1.25* is assigned to the variable *DO15I*. In FORTRAN, spaces are not significant.) In both of these statements, the lexical analyzer has to look ahead until it encounters either the "," or the "." before it is clear what the sequence of tokens is. In the first case we have the tokens "DO", "15", "I", "=", "1", "," and "25". In the second case, we have the tokens "DO15I", "=" and "1.25". In neither case can we be sure of the tokens until after the (second) "1".

Finding tokens is easier in modern languages. Each of the following streams of characters contains tokens which can be found without backtracking.

```
IF (A = B) THEN ...          Pascal, FORTRAN, ...
DECLARE (A1, A2, ...An)      PL/I
x+y                          Pascal, PL/I, C, Ada,...
x++                          C
```

One reason for this is that words such as *IF* and *DECLARE* (found in some high-level languages) are restricted as reserved, that is, they cannot be used as identifiers.

We still need look ahead to find some two-element tokens, e.g., ":=" vs. ":", or to find the ends of some tokens, e.g., "3.14" vs. "3..14". Both of these two-element tokens (:= and ..) exist in Pascal. Special tokens called *delimiters* mark the end of variable length constructs in many of today's languages. For example, ";" is used to terminate statements in Ada.

EXAMPLE 1   Again from FORTRAN

The following is the beginning of legal statements in FORTRAN:

```
IF (1.E+6 .LT. N) ...
```

and

```
IF (1 .EQ. N) ...
```

(For readers who do not know FORTRAN, the first fragment is read "If 1×10^6 is less than N", while the second is read "If 1 is equal to N...".)

The problem in Example 1 is that, having scanned the first six characters, we do not know whether

```
IF (1.E ...
```

is the beginning of

```
IF (1.E+6 .LT. N) ...
```

or

```
IF (1 .EQ. N) ...
```

In the second case, "1" is a separate token, while in the first case, it is part of the token "1.E+6". We do not know this until we have seen the "Q" in ".EQ." in the second case and the "+" in "1.E+6" in the first. Writing a program to recognize such tokens is tricky and involves looking ahead and backtracking.

A good language design minimizes such looking ahead and backtracking.

Sometimes lexical analyzers read and find tokens line by line; sometimes they read whole statements (any of which may span several lines). Often, however, they just recognize the next token when asked to do so by a calling procedure.

2.4   Major Functions of Lexical Analyzers

There are four major functions of a lexical analyzer as it finds tokens: (1) character and line handling utilities, (2) predicate testing, (3) actions and (4) error handling.

2.4.1   Character and Line Handling Utilities

A lexical analysis view of the world is portrayed in Figure 1.

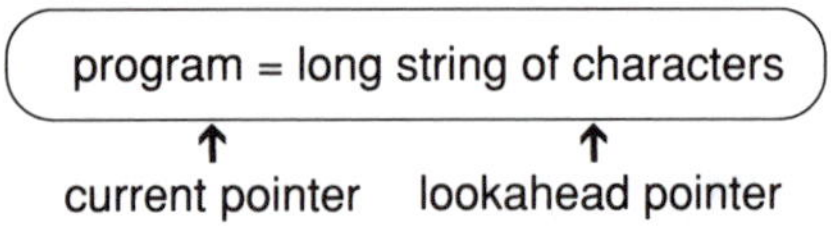

Figure 1   Lex View of the World

In Figure 1, the operations performed using the two pointers are:

GetChar—Move lookahead pointer and return next character
Fail—Move lookahead pointer back to current pointer
Retract—Move lookahead pointer back one character
Accept—Move current pointer ahead to the lookahead pointer

We will use these operations in Section 2.5 when we show how a lexical analyzer might be implemented.

In addition to the above utilities there are internal issues such as *blank suppression*. Elimination of blanks may be accomplished by (1) reading a line into an internal buffer, (2) copying the line with blanks removed into a new buffer, and (3) adding an end-of-line character.

Blank elimination might be implemented as:

```
WHILE blank DO GetChar;
```

where the operation GetChar sets the Boolean *blank* to false when it finds a non-blank.

Analogous steps could be taken if processing is done by statement rather than by line.

To put the program in a uniform and compact format, the lexical analyzer may eliminate other unneeded information such as comments. It also processes compiler control directives (such as the request to create a listing file of the program), enters preliminary information (such as the user-defined names) into tables and formats and lists the program.

2.4.2 Predicate Testing

Predicate testing checks for membership in a set of characters. Typical utilities are:

```
IsLetter  IsDigit  IsDelimiter  IsC
```

where *IsLetter(x)* is a predicate which returns true if x is a letter, *IsDigit(x)* is a predicate which returns true if x is a digit, etc. Implementations of these depend on the language in which the lexical analyzer is written. For example in Pascal:

```
IsC(x):     x = "C"

IsLetter(x):    "A" ≤ x and x ≤ "Z"
                     or
              x in ["A" .. "Z"]
```

The predicate *IsLetter* here tests for uppercase only (and presumes all uppercase letters are encoded consecutively).

2.4.3 Actions

In general, there is one action for each token type. For example,

> *InstallName:* Put name token into a name table
> *InstallLiteral*: Put literal token into a literal table

In some cases, we may want an action for each character scanned. For example, when scanning the digits in a number, we may wish to convert the sequence of characters to a number. We will see this in Section 2.5 when we scan for an integer.

2.4.4 Errors

Certain errors can be detected during lexical analysis. If the language limits the size of identifiers, this can be reported; illegal characters not belonging to the language can be detected. Anything which the lexical analyzer cannot recognize and classify is an error.

For example, most versions of Pascal require a "0" before a leading decimal point in floating point numbers: "0.5" is a legal token while ".5" is not.

Some errors, such as omitting semicolons in languages which require them or forgetting to match an open parenthesis with a closed parenthesis, cannot be found by the lexical analyzer. It merely reads sequences of characters and decides whether the word (token) is in the language or not.

2.5   Transition Diagrams

Transition diagrams describe the actions needed to recognize a token. Formally, a transition diagram is a *directed graph* with *labeled arcs*. The nodes are called *states* and the arcs are labeled with *input characters* indicating input characters which can occur before and after that state. Often a double circle is used to represent a *final* or *accepting* state. Following a transition diagram from an initial state to a final state confirms that the characters in the token are exactly those for which the transition diagram was designed.

The transition diagrams assume we are at the first character of a token, so in any implementation, the first state should skip blanks to the first non-blank.

We will show a series of examples of transition diagrams, followed by pseudo-code showing how the transition diagram might be implemented.

EXAMPLE 2   *Identifier* An identifier is described as a letter followed by an arbitrary number of letters and digits

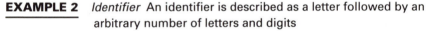
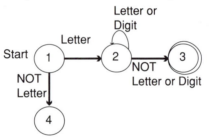

Here state 1 is the starting point for processing a sequence of characters. The label "Letter" between state 1 and state 2 means that in order to "go to" state 2, the sequence must begin with a letter. If the sequence begins with something other than a letter, state 4 is entered. This is a "failure" state: it fails the test to be an identifier.

Once state 2 is entered, processing stays in state 2 as long as the next character is either a letter or a digit. As soon as a character is not a letter or a digit, state 3 is entered. Entering this state indicates that an identifier has been found, but one too many characters has been processed.

This diagram can be described in pseudo-code as follows:

```
{state 1}: C := GetChar
           IF IsLetter (C) THEN
{state 2}:    WHILE IsLetter (C) or IsDigit (C) DO
                  C := GetChar
              ENDWHILE
{state 3}:    Retract — we have scanned one character too far
              t := (Id, Install)
```

```
                    Accept
                    Return (t)
        {state 4}: ELSE Fail — start looking for a different token
                   ENDIF
```

The actions *retract, accept,* and *fail* were described in Section 2.4.1. The action *install* might take all the characters from the current pointer to the lookahead pointer, enter the token *t* into the name table and return a pointer as value.

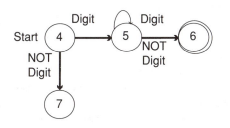

EXAMPLE 3   *Integer Constant* An integer constant is a sequence of digits

```
{state} 4 :   C:= GetChar
              IF IsDigit (C) THEN
                  Value:= Convert (C);
{state} 5:    C:= GetChar
              WHILE IsDigit (C) DO
                  Value:=10*Value + Convert (C);
                  C := GetChar
              ENDWHILE
{state} 6:    Retract ;
              t:= (Int, Value);
              Accept;
              Return (t)
{state} 7 :   ELSE Fail ; — Look for different token
              ENDIF
```

Here, the action *convert* turns a character representation of a digit into an integer in the range 0 to 9.

2.6   Finite Automata

Transition diagrams are an implementation of a formal model called *finite automata*—also called *finite-state machines* or (less often today) *sequential machines*.

Finite automata come in different flavors:

- Non-deterministic (NFA)
- Deterministic (DFA)
- Minimal deterministic

We will distinguish these by showing some examples of each.

EXAMPLE 4   Non-deterministic finite automaton

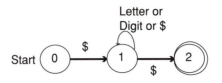

The first question we might ask is "what does this transition diagram accept?", that is, what input characters could possibly take us from the start or 0 state to the final state 2?

The answer is that this finite automaton describes special identifiers used by the author in a piece of system software she once wrote. Special identifiers, which users of the system were unlikely to use, were necessary since there could not be two identifiers with the same characters. The author chose identifiers which began with "$", contained any number of letters, digits and "$"'s, and terminated in "$". (The users of the system were told not to start their identifiers with a "$".)

The above finite automaton recognizes exactly these identifiers and no others. It is non-deterministic because in state 1, when "$" is encountered, it cannot be determined whether to stay in state 1 or proceed to state 2; that is, there are two choices for the same input.

The second question might be "can a computer program be written for it?" It may appear initially that the answer is "yes" (especially since the author just said she did!). However, *as drawn,* the answer is "no" since a computer or a computer program cannot determine upon encountering "$" whether to stay in state 1 or move to state 2 *based on no other information.*

EXAMPLE 5   Another non-deterministic finite automaton

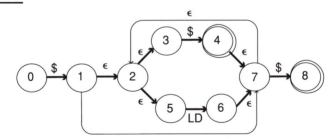

Here, LD stands for "Letter or Digit."

Our first question for the automaton in Example 5 might be: "What does $\overset{\epsilon}{\rightarrow}$ mean"?

We use $\overset{\epsilon}{\rightarrow}$ to indicate that a move can be made without processing any input, and this move is called an *epsilon transition.* Again, this may cause a non-deterministic situation as in state 7 where a decision based on no other information must be made whether to look at input and perhaps go to state 8 or look at no input and go to state 2.

The second question again might be: "What does this accept"? By careful examination, we notice that the sequence of characters must begin and end with "$". In between can be any number of letters, digits or $'s so this accepts the same sequence of characters as in the preceding example.

The third question "can a computer program be written for it?" must be answered "no" since (as shown) no computer can decide *based on no other information* whether to look at input characters or not.

EXAMPLE 6 Deterministic Finite Automaton (DFA)

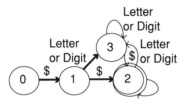

This accepts the same sequence of characters as the preceding two examples—that is, sequences of letters, digits and "$" which begin and end in "$".

Because it can be definitely determined what to do for each input character, this is a deterministic finite automata. A purist would require a deterministic automaton to have a transition from every state on every possible input. In this case, we would have a transition from state 0 when the input is a letter or digit to an error state.

A computer program can easily be written to simulate this finite automaton.

By now, the reader is probably convinced that deterministic finite automata are more practical than non-deterministic ones. However, it is often easier to "think up" non-deterministic finite automata. Then, we can "paste" transition diagrams together by using epsilon transitions. Example 7 shows the transition diagrams of Examples 2 and 3 put together with initial epsilon transitions. This finite automaton accepts either an identifier or an (unsigned) integer.

EXAMPLE 7 Non-deterministic finite automaton which accepts an identifier or an unsigned integer

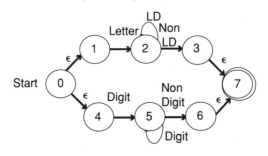

Thus to describe the tokens for a programming language, we need only describe each token and piece the transitions together. To do this automatically will result in ε-transitions. We could probably see how to eliminate the ε-transitions informally, but a computer will need a more formal mechanism to accomplish this. Exercise 13 describes an algorithm for converting from non-deterministic finite automata to deterministic finite automata, eliminating ε-transitions.

2.7   Basics from Language Theory

The finite automata described in Section 2.6 are a formalism for describing sets of strings. They are used in compilers for recognizing the tokens in a programming language. There are other formalisms for describing these same sets. One of these, *regular expressions*, is described in Section 2.8. This section introduces and reviews some basic definitions used in defining regular expressions.

2.7.1   Basic Definitions

Before we discuss regular expressions, we need to learn some vocabulary; we have already seen some of the definitions below (e.g., ε and *string*), but we repeat them here.

Definition 1: alphabet: a finite set of symbols used in a language.
> For example,
>
>   A = {$,A..Z,0..9}
>
> denotes an alphabet called *A* consisting of the symbols *$* and the set of (uppercase) letters and digits. (We have used the shorthand A..Z and 0..9, rather than listing out all the letters and digits.)

Definition 2: string = sentence = word: a finite sequence of symbols from an alphabet.
> For example,
>
>   x = $AB1$
>
> denotes a string called *x* beginning and ending with "*$*" and containing the characters "*AB1*" between the "*$*"s.

Definition 3: | *x* |: length of string *x*.
> For example,
>
> |$AB1$| = 5

Definition 4: ε: empty string.
>
> |ε| =0

Definition 5: xy: string concatenation of the strings *x* and *y*.
> For example,
>
>   if *x* = $AB1$ and *y* = $$, then *xy* = $AB1$$$
>
> Also,
>
>   xε = εx = x

Definition 6: language: a set of strings.
For example,

$$L_1 = \{\$\$,\ \$AB1\$\}$$

denotes a language L_1 consisting of the two strings *$$* and *$AB1$*.
Another example is:

```
L₂ = {$x $ | x is a sequence of letters or digits
        or $'s}
```

Here, L_2 denotes the language of Sections 2.5 and 2.6, strings of letters and digits and dollar signs, $, surrounded by dollar signs. Example 5 shows one of the finite automata which accepts this language. If we call this automaton M, then

$$L_2 = \mathscr{L}(M)$$

says that L_2 is the language accepted by M.
If we use { } to denote the empty language and {ε} to denote the language containing the empty string, then

$$\{\ \} \neq \{\epsilon\}$$

That is, these two sets are different.

Definition 7: LM: concatenation of languages L and M = {xy | x ε L, y ε M}.
For example,

```
If L = {$} and M = {x | x is a sequence of letters
           or digits}
```
then

```
LM = {$x|x is a sequence of letters or digits}
```

that is, LM consists of strings of letters and digits beginning with a dollar sign, $.

Definition 8: L^i: L concatenated i times.
For example, if L = {x | x is a sequence of letters or digits}, then L^3 = {xyz | x,y, and z are sequences of letters or digits}. L^3 consists of sequences of letters and digits of at least length three. L^0 is defined to be the language consisting of ε, that is,

$$L^0 = \{\epsilon\}$$

Definition 9: L ∪ M: {x | x ε L or x ε M}.
For example, if L = {$} and M = {$AB1$}, then L ∪ M = {$, $AB1$}.
The union of the empty language with a language L is L:

$$\{\ \} \cup L = L \cup \{\ \} = L$$

The union of {ε} with L may not equal L, however, unless L contains ε:

$$\{\epsilon\} \cup L = L \cup \{\epsilon\} \neq L \text{ (unless } \epsilon \text{ is in L)}$$

Definition 10: *Kleene closure of L,* $L^* = L^0 \cup L^1 \cup L^2 \cup \ldots$
  For example, if L = {x | x is a letter or digit}, then L^* = {sequences of letters or digits} $\cup$ {ϵ}.

Definition 11: $L^+ = L\,L^* = L^1 \cup L^2 \cup L^3 \cup \ldots$
  For example, if L = {x | x is a letter or digit}, then L^+ = {sequences of letters or digits}.

Now that we have some definitions and notation, we are ready to look at regular expressions.

2.8   Regular Expressions

A regular expression is a formula for denoting "certain" languages. Notice that we said *language,* not *string*. A single regular expression denotes an entire *set of* strings, i.e., a *language*, not a single string. Not all languages can be expressed using a regular expression. The following table recursively defines what languages can be described by regular expressions:

Regular Expression	Denotes	
ϵ	{ϵ}	
a	{a}	where a is in the alphabet
$R_1	R_2$	$L_{R_1} \cup L_{R_2}$	where R_1 and R_2 are regular expressions for L_{R_1} and L_{R_2}
$R_1 \bullet R_2$ (or $R_1 R_2$)	$L_{R_1} L_{R_2}$	
R^*	L_R^*	
(R)	L_{R_1}	

As the table shows, the language consisting of the empty string is denoted by the regular expression ϵ.
  Given an element of an alphabet, say *a*, then regular expression *a* denotes {*a*}, i.e., *a* denotes the language containing only the symbol *a*.
  The last entry allows us to use parentheses for grouping although there is a precedence order—* takes precedence over •, and • takes precedence over |, in the same way that, in arithmetic, exponentiation takes precedence over multiplication and multiplication takes precedence over addition. Notice that the concatenation operator, •, is sometimes omitted.
  Using the union, concatenation, and * operators, possibly larger languages can be expressed. Thus, if our alphabet contains {*a, b, c*}, then regular expression *a* denotes {*a*}, *b* denotes {*b*}, etc. Then:

  a | b denotes {*a, b*}, that is, the language consisting of *a* and *b*.

  ab denotes the language {*ab*}, that is, the language consisting of the string *ab*.

a^* denotes the language $\{\epsilon, a, aa, aaa, aaaa, ...\}$.

$a \mid bc$ denotes the language $\{a , bc\}$.

EXAMPLE 8   Regular expression for the language of Examples
4, 5, and 6

If our alphabet is {$} U LD where LD is a letter or digit, that is:

```
LD = {0|1|2...|9|A|B...|Z|a|b|...z}
```

then

```
$ (LD | $)*$
```

denotes the language which begins and ends with *$* and contains any
number of letters and digits and dollar signs in between.

The inputs to lexical analyzer generators, also called scanner generators, are regular expressions. These are discussed in Section 2.10.

Regular expressions can easily be converted to finite automata as described in the next section.

2.9   Converting Regular Expressions to Non-deterministic Finite Automata

Regular expressions and finite automata are alternative notations for the same type of language; that is, they have the same expressive power. If we can denote a language by a regular expression, then we can draw a finite automaton to represent it, and vice versa. The following pictures show how a regular expression can be represented by non-deterministic finite automata.

ε:

1.
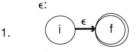

This first diagram shows that the regular expression ε which denotes the language consisting of the empty string, {ε}, can be recognized by the non-deterministic finite automaton which goes from its initial to its final state by reading no input, that is, by an ε-transition.

a:

2.
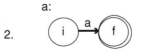

This diagram shows that the language {*a*} denoted by the regular expression *a* is recognized by the finite automaton which goes from its initial to final state by reading an *a*.

3. R₁|R₂

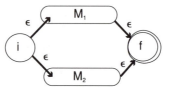

Here, we presume that the languages denoted by the regular expressions R_1 and R_2 and L_{R1} and L_{R2}, respectively, are recognized by the finite automata denoted by M_1 and M_2. Adding ϵ-transitions to the beginning and end creates a finite automaton which will recognize either the language represented by R_1 or the language represented by R_2, i.e., the union of these two languages, $L_{R1} \cup L_{R2}$. (We could eliminate the new final state and the ϵ-transitions to it by letting the final states of M_1 and M_2 remain final states.)

4. R₁ • R₂

In diagram 4, M_1 is the machine which accepts L_{R1} and M_2 is the language which accepts L_{R2}. Thus, after this combined automaton has recognized a legal string in L_{R1}, it will start looking for legal strings in L_{R2}. This is exactly the set of strings denoted by $R_1 \cdot R_2$.

5. R*

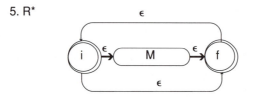

Here, M is a finite automaton recognizing L_R. Adding a new initial state which is also a final state recognizes the empty string, ϵ (if L_R didn't contain ϵ already). The ϵ-transition from all final states of M to its initial state will allow the recognition of L^i_R, i = 2, ...

Using these diagrams as guidelines, we can "build" a non-deterministic finite automaton from any regular expression.

EXAMPLE 9   Converting a regular expression to a finite automaton

Create a finite automaton for (x|y|z)*x.
We will build this in steps:

Step 1 (x|y|z)

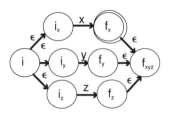

Here, the finite automata accepting {x}, {y}, and {z} have been linked together by providing ε-transitions into and out of them from new initial and final states. We actually skipped step 0 here which would have created the three finite automata accepting {x}, {y}, and {z}. The notation f_{xyz} denotes the final state of an automaton which accepts the language denoted by the regular expression x|y|z.

Step 2 (x|y|z)*

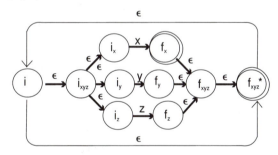

To recognize (x|y|z)*, we have added new initial and final states and the appropriate ε-transitions. We could actually have used the old initial and final states, adding the two ε-transitions. Here, $f_{xyz}*$ denotes the final state of an automaton which accepts the language (x|y|z)*.

Step 3 (x|y|z)*x

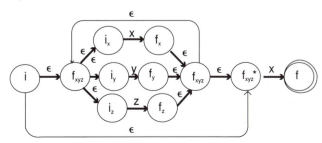

A new transition has been added that takes us to a new final state. The finite automaton is complete. Of course, it is non-deterministic; the exercises show techniques for turning this into a deterministic finite automaton. Sometimes, if the regular expression to be recognized is simple, it is possible to "guess at" the deterministic finite automaton. The reader might try to create a DFA for this example.

2.10   Scanner Generators

There are many tools available today which will generate the lexical analyzer automatically from a description of the tokens of the language. Usually the input is a regular-expression-type syntax. Once the lexical analyzer is created, programs are read in and tokens are found.

The transition diagrams which implement a lexical analyzer are essentially a finite state machine. A finite state machine can be represented by a table (see Exercise 9). A lexical analyzer generator, or scanner generator, reads regular expressions representing the tokens and creates this state transition table. Then a "driver" routine reads input, checks the table and finds the tokens. Thus, a scanner generator can be broken into two parts—(1) the table generator which inputs regular expressions and outputs a table and (2) the driver routine which reads the table and the input, and outputs the tokens.

This chapter describes how to generate a lexical analyzer (as well as how to write one): Section 2.9 describes how to convert a regular expression into a non-deterministic finite automaton. Exercises 13 and 14 continue the process.

Probably the most famous scanner generator is LEX, a tool which comes with the UNIX operating system.

2.11   Summary

This chapter has described the first phase of the compiling process—lexical analysis. Lexical analysis is the process of breaking the input stream of characters into a stream of tokens, which are ultimately input to the parser. In some compiler implementations, the parser may actually motivate this call for tokens.

Lexical analyzers are implemented from transition diagrams or their formal counterpart, finite state automata. In this chapter we saw examples of transition diagrams for common programming language tokens, e.g., *identifiers* and *integers.*

Writing a lexical analyzer is a task often encountered in computer science. Lexical analyzers are an initial phase to many language processors, not just compilers. Assemblers must lexically analyze input before transforming instruction mnemonics to machine code.

Another formalism for describing tokens is a regular expression. Regular expressions denote a language, the same set of languages accepted by finite automata. (See Section 2.9 and Exercise 16.) In this chapter, we saw how to convert regular expressions to (non-)deterministic finite automata. The exercises explore how to convert (automatically) a non-deterministic finite automaton to a deterministic one. By hand, a deterministic finite automaton can often be written heuristically by trial and error.

The exercises also explore other mechanisms for describing tokens.

Lexical analyzers can also be generated automatically by generator programs. Such programs actually create a minimal finite automaton from a regular expression describing the tokens.

EXERCISES

1. For the following program written in a subset of Ada, find and classify all the tokens:

```
BEGIN
    A := B3 ;
    Xyz := A + B + C
                    - P/Q ;
    A := Xyz * (P + Q);
    P := A - Xyz - P;
END;
```

2. Show transition diagrams for the tokens listed in Exercise 1.

3. Show a regular expression for the tokens in Exercise 2.

4. Write and compile an erroneous program in your favorite language. Create two lists: (1) a list of those errors which you think were found by the lexical analyzer and (2) a list of those not found by the lexical analyzer. Justify each entry according to the functionality of a lexical analyzer.

5. Repeat Chapter 1, Exercise 2.

6. Given a language consisting of the Pascal relational operators (<, <=, =, <>, >, >=), design a deterministic finite automaton with as few states as possible, represented as a transition diagram.

7. Consider the following transition diagram:

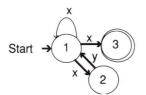

   (a) Does it represent a deterministic or a non-deterministic finite automaton?
   (b) Describe in words the language accepted here.

8. Given the following regular expressions for Pascal unsigned numbers without an exponentiation E operator:

Number = Digit$^+$ OptionalFraction
OptionalFraction = .Digit$^+$ | ϵ

    (a) Rewrite *Number* as a single regular expression using the following short-hand if needed:

        *: 0 or more occurrences

        +: 1 or more occurrences

        ?: 0 or 1 occurrence

    (b) Write a deterministic finite automaton which recognizes *Number*.

9. State Tables: This chapter represents finite automata using transition diagrams. They can also be represented using a *state table*.

    A state table lists input symbols along the top and each state down the left. Thus each row represents a state and each column represents an input symbol. Entries are the values of Table[state,input], in the same sense that A[I,J] denotes the entry in the ith row, jth column of an array A. The state table for the finite automaton of Example 4 is:

	$	Letter or digit
0	1	—
1	1,2	1
2	—	—

Show the transition diagram of Exercise 7 as a state table.

10. If $A = \{x, y, z\}$ what languages are denoted by the following regular expressions? Describe your sets in English as unambiguously as possible.

    (a) x

    (b) x | y

    (c) x y | z

    (d) (x y) | z

    (e) x (y | z)

    (f) (x | y)*

    (g) (x* | y*)

    (h) x | (yz)*

    (i) (x | y | z)*x

11. Given the alphabet $\{a, b\}$, denote the following languages by regular expressions *if you can*:

    (a) Strings of a's and b's beginning with a.

    (b) Strings of a's and b's consisting of an even number of a's.

    (c) Palindromes consisting of a's and b's, that is, all strings which read the same backwards as forwards, e.g.,

        abbbba   abaaba   bbbabbb, etc.

    (d) Strings which end in ab and which are at least four characters long.

12. For $L = \{a\}$, show L^3.

13. Converting from an NFA to a DFA: In Section 2.6, we described a set of strings using three different finite automata, two non-deterministic and one deterministic. We can always convert a non-deterministic finite automaton to a deterministic one. Using the automaton in Example 5, we do this by:

(1) *Grouping* states of the old finite automaton. A new state in the new automaton will consist of a **set of** states from the old automaton. That is, sets of states of the old automaton will denote a single state in the new automaton.

(a) The new start state contains one member—the old start state.

(b) For an input, say, *a*, the next state will contain all states accessible from the previous state upon scanning an *a* in the input. Thus for the automaton of Example 5, the new start state 0' would be {0}. The new state 1' = {1,2,3,5,7} since from state 0' we can access all these states upon scanning a "$". (Try it.)

(c) Repeat step (b) for each newly added (and unprocessed) state until no new states are introduced.

(2) Final or accepting states are any new states which contain a final state from the original non-deterministic automaton. Thus in our example, any state containing either 4 or 8 will be a final state.

One method for beginning the conversion from a non-deterministic finite automaton to a deterministic one is to write a little preprocessor which lists all the states accessible from a state via ε-transitions. Then it is easier to see what states are accessible upon scanning the input from a particular state—they are the states accessible upon scanning the symbol plus those accessible by ε-transitions. We show this below.

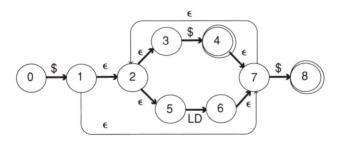

Preprocessor

state	ε-transition state
0	0
1	1,2,3,5,7
2	2,3,5
3	3
4	4,7,2,3,5
5	5
6	6,7,2,3,5
7	7,2,3,5
8	8

Note that we can always make an ε-transition to the same state. Now, we use the preprocessor to make it easier to see the transitions on our two inputs: $ and LD, where, again, LD stands for a letter or a digit:

Using steps (1) and (2) described above,

state	$	LD
0	1' = {1,2, 3,5,7}	–
1	2' = {4,7, 2,3,5,8}	3' = {6,7 2,3,5}
2	{8,4,7,2, 3,5} = 2'	{6,7,2,3, 5} = 3'
3	{8,4,7,2, 3,5,} = 2'	{6,7,2, 3,5} = 3'

New States:

0' = {0}
1' = {1,2,3,5,7}
2' = {2,3,4,5,7,8}
3' = {2,3,5,6,7}

We can draw this new DFA as a transition diagram using the states 0', 1', 2', and 3':

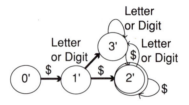

This is the same finite automaton as in Example 6. This automaton is not minimal, that is, there is a finite automaton with fewer states which accepts the same identifiers. The method described to convert from a non-deterministic finite automaton to a deterministic one does not guarantee that the resulting automaton is minimal. Use this method to create a deterministic finite automaton for the diagram in Exercise 7.

Exercise 14 describes how to turn a non-minimal deterministic automaton into a minimal one.

14. **Finding a Minimal DFA:**   We can minimize a DFA by looking for two states which represent *exactly the same situation* and then merging them into one state. The method is called *partitioning* and we start by creating two groups— final states and all others. We continue by separating out any states which lead to states in different partitions.

Consider the following DFA which accepts the same set of strings as in the preceding examples (identifiers beginning and ending in "$", but with any number of letters, digits, and "$" in between). We know this can be done with fewer than six states. (See Example 6 and Exercise 13.)

Studying this automaton reveals that states 3 and 5 and states 2 and 4 "seem" to be the same. States 3 and 5 both lead to 4 on input $; they lead to each other on input LD and both are accessed by reading input LD. Our minimiza-

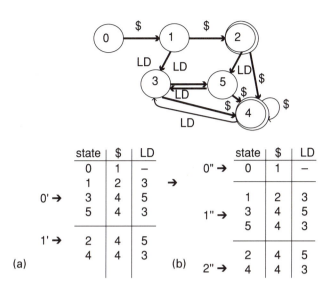

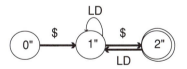

tion algorithm will discover this. The automaton has been written in table form above and the table labeled (a) shows the two final states 2 and 4 separated out.

Looking at the upper set of states in (a), that is, in state 0', we see that states 1, 3, and 5 on input $ lead to states in the bottom set, that is, they lead to state 1'; thus, on input $, states 1, 3, and 5 do the same thing. State 0' does not lead to a state in the bottom set on input $. On input LD, states 1, 3, and 5 lead to 3, or 5 again, i.e., to states in 0'. Thus, we separate states 1, 3, and 5 from state 0.

We are finished since state 0" has only one member; states 1, 3, and 5 in 1" do the same thing (lead to state 2" on a "$" and to state 1" on LD); and finally, states 2 and 4 in 2" also do the same thing (lead to 2" when the input is "$" and lead to state 1" when the input is "LD").

Drawing out the finite automaton, we have:

Following these steps , determine if the DFA of Example 6 is minimal; if not, convert it to a minimal DFA.

15. Convert the automaton of Exercise 7 to a minimal one.

16. Regular expressions and finite automata denote the same set of languages. We proved the easy half of this in Section 2.9, where we showed how to represent regular expressions with finite automata. More difficult is the following: Show how the languages accepted by finite automata can be represented as regular expressions.

Compiler Project Part I

Lexical Analysis

The following are tokens in a very small subset of Ada:

```
begin
end
;
:=
(
)
+
-
*
/
mod
rem
identifiers
integers
```

where identifiers and integers are as defined in Examples 2 and 3, respectively.

Decide on classifications for the above tokens and write a lexical analyzer which will find and classify the tokens. Different classifications for + and − (maybe call them AddOps) and for *, /, *mod* and *rem* (maybe call them MulOps) will make life easier later in the parser.

Run your lexical analyzer on (a) the following program and (b) a program of your own design.

```
begin
  a := b3;
  xyz := a + b + c
              - p / q;
  a := xyz * (p + q);
  p := a - xyz - p;
end;
```

Your output should show the input lexeme and its type as in the following example:

Type	Lexeme
Keyword	*begin*
Identifier	*a*
AssignOp	*:=*
(etc.)	

(You may choose to classify *begin* as an identifier.)

An extension to this would be to create a name table which stores identifiers in a table using techniques from data structures to create the table. After the tokens are output, output the entries in the name table in alphabetical order (if you use a hashing algorithm for your table, you will need to sort):

Names
a
b
b3
(etc.)

Alternate Assignment 1 (Easier): Use a lexical analyzer generator

Using a scanner generator of your choice, create a scanner. Use it to find the tokens as in the regular assignment above.

Alternate Assignment 2 (Much, Much Harder): Create a lexical analyzer generator

Write a scanner generator in two parts—(1) a (transition) table generator, which inputs regular expressions for the tokens and outputs a state transition table as in Exercise 9, and (2) a driver routine which reads the table and the input and outputs the tokens.

3

Introduction to Grammars and Parsing

Time flies like an arrow;
fruit flies like a banana.
—Groucho Marx

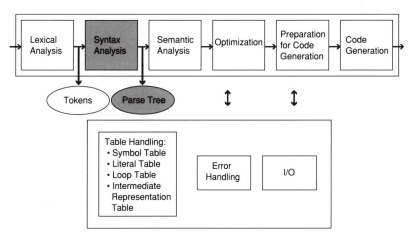

3.0 Introduction

Parsing, also called *syntax analysis*, is the phase which tries to group sequences of tokens into syntactic categories. If the parser cannot find such a grouping, then it reports that a *syntax error* has occurred.

EXAMPLE 1   Characters, tokens and syntactic categories

Consider the sequence:

```
4 * Max
```

This consists of seven characters (counting embedded spaces). A lexical analyzer might find three tokens (*4*, ***, and *Max*). The syntax analyzer (parser), however, might group this all into one syntactic category, *expression*.

Many compilers are *parser-driven* which means that the parser calls the scanner whenever it needs a new token rather than receiving a complete stream of tokens on which to work.

Parsers are written to recognize syntactically correct programs in some language. Traditionally, the syntax of a language is defined by a *grammar,* that is, a set of rules defining what a legal program looks like.

Another way to define the syntax of a language is by an *acceptor* or *recognizer* which decides whether a particular string belongs to a language.

For finding tokens, acceptors called finite automata were used (in Chapter 2) although another formalism, regular expressions, was discussed. Regular expressions are useful in the automatic generation of lexical analyzers.

In parsing, *context-free grammars* are the most useful formalism and we will discuss them first.

3.1 Grammars

Grammars describe languages. Natural languages such as English are often described by a grammar which groups words into syntactic categories such as subjects, predicates, prepositional phrases, etc.

Stated more mathematically, a grammar is a formal device for specifying a potentially infinite language in a finite way, since it is impossible to list all the strings in a language whether it is English or Pascal. At the same time, a grammar imposes a structure on the sentences in the language. That is, a grammar, G, defines a language L(G) by defining a way to derive all strings in the language. We will look at this first for a (very) small subset of English.

3.1.1 Context-free Grammar for English

Noam Chomsky (1957), in 1957, used the following notation, called *productions,* to define the syntax of English. The terms *sentence*, *noun phrase,* etc. plus the following rules describe a small set of English sentences. The articles *a* and *the* have been described, here, as adjectives for simplicity.

\<sentence\>	→	\<noun phrase\>\<verb phrase\>
\<noun phrase\>	→	\<adjective\>\<noun phrase\> I \<adjective\>\<singular noun\>
\<verb phrase\>	→	\<singular verb\>\<adverb\>
\<adjective\>	→	A I The I little
\<singular noun\>	→	boy
\<singular verb\>	→	ran
\<adverb\>	→	quickly

Here, the arrow, →, might be read as "is defined as" and the vertical bar, "I", as "or". Thus, a noun phrase is defined as an adjective followed by another noun phrase or as an adjective followed by a singular noun. This definition of noun phrase is recursive because \<noun phrase\> occurs on both sides of the production. Grammars are often recursive to allow for infinite length strings.

This grammar is said to be *context-free* because only one syntactic category, e.g., *\<verb phrase\>*, occurs to the left of the arrow. If there were more than one syntactic category this would describe a context and be called *context sensitive.*

A grammar is an example of a *metalanguage*—a language used to describe another language. Here, the metalanguage is the context-free grammar used to describe a part of the English language.

EXAMPLE 2   Finding the structure of a sentence

Using the grammar above, the sentence:

The little boy ran quickly

can be diagramed :

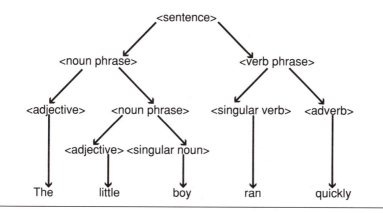

Most people would agree that "Quickly, the little boy ran" is also a syntactically correct sentence, but it cannot be derived from the above grammar. In fact, it is impossible to describe all the correct English sentences using a context-free grammar.

On the other hand, it is possible, using the grammar above, to derive the syntactically correct, but semantically incorrect, string "Little the boy ran quickly".

3.1.2   Context-free Grammars for Programming Languages

The structure of English is given in terms of subjects, verbs, etc. The structure of a computer program is given in terms of procedures, statements, expressions, etc. For example, an arithmetic expression consisting of just addition and multiplication may be described using the following rules:

```
<expression>::=  <expression> + <term> | <term>
<term>       ::=  <term> * <factor> | <factor>
<factor>     ::=  (<expression>) | <name> | <integer>
<name>       ::=  <letter> | <name> <letter> | <name> <digit>
<integer>    ::=  <digit> | <integer> <digit>
<letter>     ::=  A | B | .... | Z
<digit>      ::=  0 | 1 | 2 | 3 | 4 | ... | 9
```

Here, we have used "::=" for *is defined as* rather than an arrow, →, as before. The metalanguage BNF (Backus-Naur form) is a way of specifying context-free languages and it was originally defined using ::= rather than →. Credit is usually given to John Backus and Peter Naur for developing this notation as a way of describing the syntax of the programming language Algol. As long as we understand what is meant and what the capabilities of this grammatical description are, the notation doesn't matter.

Unlike natural languages like English, all the legal strings in a programming language **can** be specified using a context-free grammar. However, grammars for programming languages specify semantically incorrect strings just as well as do grammars for natural languages. For example, context-free grammars cannot be used to tell if a variable, say, *A*, declared to be **Boolean** is used in an arithmetic expression like *A* + *1*. Thus the syntax analysis phase cannot find this error. (The next phase, semantic analysis, performs this check.)

EXAMPLE 3   Finding the structure of an expression

Using the grammar above, the string

```
A * B
```

can be diagramed:

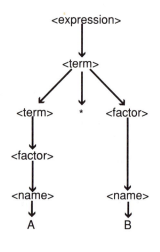

Parse Tree

The diagram above is called a *parse tree*. It shows how a string is *derived* (generated) using a grammar. Other words for parse tree are *syntax tree* and *derivation tree*.

Derivations

The parse tree above shows the structure of *A* ∗ *B*, but it does not tell us in exactly what order the rules listed at the beginning of this section were applied. (We will be concerned with the order in which the rules are applied for parsing.) Example 4 shows a simplified version of this grammar and beside it a derivation of *a* + *a* ∗ *a*. We will use this grammar to define some notation needed to describe how a string is parsed.

EXAMPLE 4   The (simplified) expression grammar and a derivation

Grammar   **Derivation of** $a + a * a$

1. E → E + T $E \xrightarrow{1} E + T \xrightarrow{2} T + T \xrightarrow{3} T + T * F$
2. E → T
3. T → T * F $\xrightarrow{4} F + T * F \xrightarrow{6} F + T * a$
4. T → F
5. F → (E) $\xrightarrow{6} a + T * a \xrightarrow{4} a + F * a$
6. F → a
                             $\xrightarrow{6} a + a * a$

In Example 4, E stands for *expression,* T for *term,* etc. The terminal a represents any legal operand in an arithmetic expression. In the derivation, the number above each arrow is the number of the production applied.

Left and Right Derivation

It may seem, in Example 4, that the rules are applied in a haphazard order (they are). In parsing, rules are often applied by continually replacing the left-most or the right-most symbol. These two derivation orders are called, respectively, left and right derivations.

EXAMPLE 5   Right derivation of $a + a * a$

$E \xrightarrow{1} E + T \xrightarrow{3} E + T * F \xrightarrow{6} E + T * a$

$\xrightarrow{4} E + F * a \xrightarrow{6} E + a * a$

$\xrightarrow{2} T + a * a \xrightarrow{4} F + a * a$

$\xrightarrow{6} a + a * a$

The rules applied are:
1 3 6 4 6 2 4 6
If we apply these rules in the order listed, we can draw a parse tree for this string. Thus, the list of rules applied represents a parse tree.

Terminals

A grammar consists of a set of terminal symbols (tokens) such as the plus sign, +, the times sign,*, left and right parentheses, (and), and the identifier a. A terminal is a symbol which does not appear on the left-hand side of any production. (Terminals do not have to consist of only one character.)

Nonterminals

Nonterminals are the non-leaf nodes in a parse tree. In the grammar above, E, T, and F are nonterminals. In our first, more formal version of this grammar, all the nonterminals were enclosed in angle brackets, < >.

Productions

Productions were introduced in Section 3.1.1. Productions can be thought of as a set of replacement rules (also called *rewriting rules*). Each rule may be written:

      A ::= α

or

      A → α

where A is a nonterminal and α is a string of terminals and nonterminals. In the above grammar E → E + T is a production.

Start Symbol

The *start* symbol, also called a *goal* symbol, is a special nonterminal designated as the one from which all strings are derived. In our grammar, E (standing for expression) is the designated start symbol.

Sentential Form

A *sentential form* is any string derivable from the start symbol. Thus, in the above derivation of $a + a * a$, E + T * F, E + F * a, and F + a * a are all sentential forms as is $a + a * a$ itself.

Sentence

A *sentence* is a sentential form consisting only of terminals such as $a + a * a$.
    A sentence can be derived (created) using the following algorithm:

Algorithm
Derive String

```
String := Start symbol
REPEAT
  Choose any nonterminal in String.
  Find a production with this nonterminal on the
    left-hand side.
  Replace the nonterminal with one of the options on
    the right-hand side of the production.
UNTIL String contains only terminals.
```

The derivation of Example 4 uses this algorithm to derive $a + a * a$ from the expression grammar.

Extended BNF

Recursive procedures in programming can be rewritten using iteration (and a stack). Similarly, we rewrite recursive productions using iteration. Braces, { }, are often used to represent 0 or more occurrences of items on the left-hand side of a rule and brackets, [ ], to represent optional items. Thus, using extended BNF, we can write the above "expression grammar" as:

```
E → T {+ T}
T → F {* F}
F → (E) | a
```

The first rule here derives the sentential forms T, $T+T$, $T+T+T$, etc.

Of course, since F can derive an E in F $\rightarrow$ (E), this grammar is still indirectly recursive.

Epsilon Productions

A production can have a right-hand side containing just the empty string, ϵ. The following grammar is another version of the so-called "expression" grammar:

```
E  → T E'
E' → + T E' | ε
T  → F T'
T' → * F T' | ε
F  → (E)  | a
```

Here, two new symbols, E' and T', have been introduced.

Relationship of Grammars to Parsing

In some sense, parsing reverses the derivation process in that we have an input string and have to "discover" the parse tree (if any) for it.

3.2 Ambiguity

If an English sentence has more than one meaning, it is said to be ambiguous. Often such sentences can be parsed more than one way. The sentence,

> Time flies like an arrow

can be interpreted with *time* as a noun, *flies* as a verb, and *like an arrow* as an adverbial phrase. This interpretation is a comment on the fast passage of time. However, if *time* is interpreted as an adjective, *flies* as a noun, *like* as a verb, and *arrow* as a direct object noun, the sentence becomes a comment on the love life of some species called a *time fly*. There are other interpretations of this sentence based on its syntax (see Exercise 4).

Similarly, meaning is assigned to programming language constructs based on their syntax. We prefer, therefore, that programming language grammars describe programs unambiguously.

A sentence is *ambiguous* if there is more than one distinct derivation. If a sentence is ambiguous, the parse tree is not unique; we can create more than one parse tree for the same sentence.

A grammar is *ambiguous* if it can generate even one ambiguous sentence.

Examples 6 and 7 show common programming language constructs, (1) expressions and (2) the IF-THEN-ELSE statement using ambiguous grammars.

EXAMPLE 6   Ambiguous grammar for expressions

Consider the following expression grammar:

```
E → E + E
E → E * E
E → (E)
E → a
```

and input: *a* + *a* + *a*. We can find the following derivations:

Derivation 1:

$$E \rightarrow E + E \rightarrow a + E \rightarrow a + E + E$$
$$\rightarrow a + a + E \rightarrow a + a + a$$

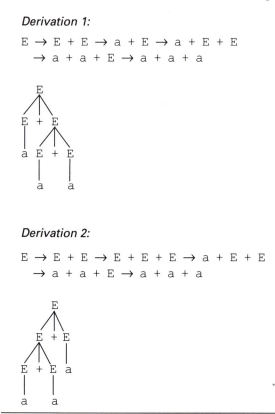

Derivation 2:

$$E \rightarrow E + E \rightarrow E + E + E \rightarrow a + E + E$$
$$\rightarrow a + a + E \rightarrow a + a + a$$

Thus, in Example 6, there are two entirely different parse trees for the same expression using the grammar. For this reason, the grammar is said to be ambiguous. In addition, it doesn't show the precedence levels as does the original grammar (E → E + T, etc.) although there are other ways to incorporate precedence.

EXAMPLE 7   Dangling else

Consider the following grammar for IF-THEN-ELSE statements:

```
S → IF b THEN S ELSE S
  | IF b THEN S
  | a
```

where *b* represents a condition and *a* represents another sequence of statements.

Consider the following string derivable from this grammar:

```
IF b THEN IF b THEN a ELSE a
```

where *b* is a condition and *a* is a statement or set of statements. This string has two parse trees:

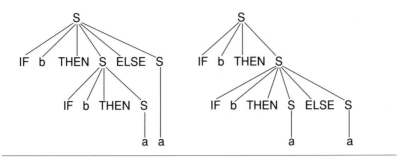

For the grammar of Example 7, the second parse is often considered the one of choice, that is, the *else* should associate with the "closest" *if*.

We can revise the grammar of Example 7 to:

```
S1→ IF b THEN S1 | IF b THEN S2 ELSE S1 | a
S2 → IF b THEN S2 ELSE S2 | a
```

Here S1 is the start symbol. Now

```
IF b THEN IF b THEN a ELSE a
```

has one parse tree:

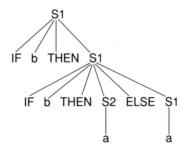

Ambiguity is not always an inherent property of languages; there exist both ambiguous and unambiguous grammars for some constructs. We know that expressions can be described unambiguously since we have done so in previous examples.

Other Language Problems

Ambiguity is not the only problem encountered in grammars. Other properties of grammars make it difficult to parse using particular techniques. For example, the expression grammar:

```
1. E → E + T
2. E → T
3. T → T * F
4. T → F
5. F → (E)
6. F → a
```

is said to be *left recursive*. E can generate a sentential form with E as the leftmost symbol. As we will see, this is a problem for parsers called *top-down* parsers.

3.3   The Parsing Problem

Briefly stated the parsing problem is to take a *string of symbols* in a language and a *grammar* for that language and, from them, to construct the *parse tree* or report that the sentence is syntactically incorrect. For correct strings:

```
Sentence + Grammar → Parse Tree
```

For a compiler, a sentence is a program:

```
Program + Grammar → Parse Tree
```

Writing a grammar for a programming language is often a difficult task. In the project described at the end of this chapter, a grammar for a (very) small subset of Ada is shown. Subsequent chapters add to this grammar.

We'll look at parsers which read the input string from left-to-right looking at most one token ahead.

We will use the following grammar which describes strings of decimal digits to illustrate the different parsing methods and problems:

```
N → D | N D
D → 0 | 1 | 2 | 3 | 4 | 5 | 6 | 7 | 8 | 9
```

3.3.1   Top-down Parsing

Top-down parsing creates a left-most derivation. The steps are:

(1) Begin with *start* symbol (for our grammar, *N*) as the "root" of the parse tree. (It is interesting that the top of the tree is called the root!)

(2) At each step, replace the left-most nonterminal *V* in the current sentential form

```
xVy
```

by *u*

where there is a rule (a production)

```
V → u
```

so that we have

```
xuy
```

EXAMPLE 8   Top-down parse of 3 5
_____

Using the grammar above, the input 3 5 would be parsed top-down as:

```
N → N D → D D → 3 D → 3 5
```

The parse tree is built step-by-step:

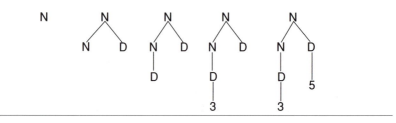

We will describe two top-down parsing methods in Chapter 4.

3.3.2 Bottom-up Parsing

Bottom-up parsing creates the reverse of a *right* derivation.

The steps are:

(1) Begin with the string (i.e., leaves of the to-be-created parse tree).

(2) Try to *reduce* to the *start* symbol by finding the current *handle*:

The handle is

(i) the largest collection of terminals and nonterminals in the leftmost part of the input which can be found on the right-hand side of some production and

(ii) such that all the symbols to the right of the handle are terminals and

(iii) such that replacing the handle with the left-hand side of the production eventually (by finding more handles) leads back to the *start* symbol.

EXAMPLE 9 Bottom-up parse of 3 5

The string **3 5** would be parsed bottom-up as:

3 5 ← D 5 ← N 5 ← N D ← N

The steps to building the tree are:

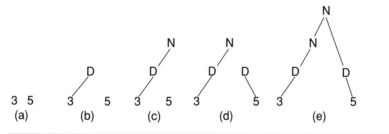

(a)   (b)   (c)   (d)   (e)

In the figure of Example 9, labeled (a), 3 is the handle since all symbols to its right (the 5) are terminals and 3 is the longest string found on the right-hand side of a production. In (b), D is the handle; in (c), 5 is the handle and in (d), N D is the handle. The reader is encouraged to justify the choice of handle in each case (see Exercise 5).

3.3.3 Parsing Issues

Both top-down and bottom-up parsing involve issues to be resolved.

Top-down Problems

(1) More than One Choice Suppose the grammar contains two options for non-terminal V:

V → u1 | u2

Which one should be picked?

One possible solution is to pick at random; remember the "state of the parse" and what is chosen. If wrong, the parser could go back and try the other choice. This is called back-up parsing or backtracking and is not very efficient. A better but more complicated way is to check the next input token(s) in the string and match it with the first terminal(s) derived from u1 or u2. Although more complicated, this is actually more efficient.

(2) Left Recursion   Consider the standard expression grammar:

```
E → E + T | T
T → T * F | F
F → (E) | a
```

and suppose we want to parse $a + a + a + ...$ without looking more than one or two tokens ahead. (In practice, parsers look at most one token ahead.) Without knowing how many "+"'s there are in the string, the parser doesn't know how many times to apply the rule E → E + T. The "+" generated by the first E → E + T might be 100 tokens away from the beginning of the string(!):

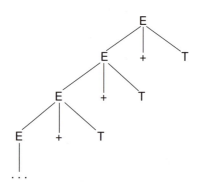

That is, without knowing how many "+"'s there are in the string (and it is not practical to look at the entire string to find out!), a top-down parser doesn't know how many times to use the production E → E + T before using the rule E → T.

If top-down parsing is employed, left recursive grammars must be changed (see Exercises 9, 10 and 11).

(3) Left-Factoring   Sometimes two productions share common symbol(s) for the beginning of their right-hand sides. This is a problem for top-down parsers which must expand one production or the other.

For example,

```
Statement → IF C THEN S ELSE S | IF C THEN S
```

contains the same first tokens, *IF C THEN S* (*C* and *S* represent many tokens).

If we change this to:

```
S  → IF C THEN S S'
S' → ELSE S | ε
```

then a top-down parser has no problem. There is a formal method for left-factoring a grammar (see Exercise 12).

Bottom-up Issues

Bottom-up parsing does not raise as many problems as top-down parsing. The first problem for bottom-up parsing is analogous to a problem for top-down parsing.

Suppose the grammar contains both productions:

```
U → ω
V → ω
```

and ω is the handle in the string being parsed; do you reduce ω to U or to V? As with the top-down case, the parse can make a random choice and backtrack if wrong. Once again, this is inefficient and a method of looking ahead is usually employed. For bottom-up parsing, however, this is more difficult. We will discuss this in Chapter 5. The interested reader might try Exercise 13.

Non-Context-Free Issues

Some problems which we might describe as syntax related cannot even be described using context-free grammars. For example, it is not possible to express the fact that an identifier declared to be of type Boolean is used only in Boolean expressions, nor even that it has been declared before it is used. A similar problem is that of matching the number and type of arguments in a procedure call to the ones declared in the procedure definition. Although these issues appear to be "syntactic", they are resolved in the semantic analysis phase of a compiler.

3.4   Parser Generators

The next two chapters describe details of various parsing methods and algorithms. However, it is possible to avoid writing these programs by using one of the many available parser generator programs.

Parser generators, like scanner generators, often come in two parts: a table generator which reads in the BNF and creates a table and a driver program which reads the input, consults the table and creates the parse tree. The tables and driver vary according to whether the parsing method is top-down (described in Chapter 4) or bottom-up (described in Chapter 5). Many of today's parser generators embed the table and the driver into a program, making the resulting tool a bit easier to customize.

A well known parser generator program is YACC, *Yet Another Compiler Compiler,* a tool which comes with the UNIX operating system.

3.5   Summary

This chapter introduces the problem of finding the grammatical structure of input programs. Code will ultimately be created based on this syntactic structure, so the parsing phase is an extremely important one.

The major functions of a parser are to (1) construct a parse tree for an input string or (2) report that the input string is not grammatically correct.

There are two primary parsing methods: top-down which creates a parse tree from the root down and bottom-up which creates a parse tree from the leaves up.

Languages are described using grammars. In particular, the syntax of a language is described using context-free grammars and written down using Backus-Naur form (BNF).

Context-free grammars often have to be rewritten to be able to implement them in a parser. Some difficulties include ambiguity, left recursion and the need for left factoring.

Many of the terms and concepts introduced in this chapter will be used in Chapters 4 (top-down parsing), 5 (bottom-up parsing) and 6 (syntactic error handling).

EXERCISES

1. Derive the string $a + a * a$ using (a) the grammar in Example 4, (b) the ambiguous grammar in Example 6, (c) the extended BNF grammar in Section 3.1.2 and (d) the grammar with epsilon productions of Section 3.1.2. In each case, draw as many parse trees as you can.

2. For the string of Exercise 1 and the four grammars, show (a) a right derivation and (b) a left derivation.

3. Study the first two exercises and devise an algorithm which will derive a given string automatically, looking at the next input symbol only. That is, devise an algorithm for top-down parsing.

4. Find another interpretation (see Section 3.2) for the ambiguous sentence: *Time flies like an arrow.*

5. Justify the choice of handles in steps (b), (c), and (d) of Example 9.

6. Using the "standard" expression grammar:

```
E → E + T
E → T
T → T * F
T → F
F → (E) | Id
```

identify the handle of:

   (a) a * a + a
   (b) T * F + a
   (c) T + F
   (d) E + F
   (e) E + T
   (f) E + T * F

7. Show that the grammar:

```
Number → Digit | Number Number
```

```
Digit → 0|1|2|...|9
```

is ambiguous. Change it to be unambiguous, but still generate the same set of strings.

8. Consider the following grammar:

```
S → (S) S | ε
```

(a) Describe the set of strings which it generates.
(b) Is it ambiguous?
(c) Is it left recursive?

9. Eliminating Left Recursion. Part I: There is a formal technique for eliminating left recursion from productions.

For each rule which contains a left recursive option,

```
A → A α | β
```

introduce a new nonterminal A' and rewrite the rule as

```
A  → β A'
A' → ε | α A'
```

Thus the production:

```
E → E + T | T
```

is left recursive with "E" playing the role of A, "+ T" playing the role of α, and "T" playing the role of β. Introducing the new nonterminal E', the production can be replaced by:

```
E  → T E'
E' → ε | + T E'
```

(Notice, however, that we have lost the original "left associative" structure.) Use this rule to eliminate left recursion from the following productions:
(a) T → T * F | F
(b) A → A X Y z | z
(c) StatementList → StatementList ; Statement | Statement

10. Eliminating Left Recursion. Part II: The method in Exercise 9 can be extended to more than two choices on the right-hand side. The general rule is to replace

$$A → A\alpha_1 | A\alpha_2 | A\alpha_3 | \ldots | A\alpha_n | \beta_1 | \beta_2 | \ldots | \beta_m$$

by

$$A → \beta_1 A' | \beta_2 A' | \ldots | \beta_m A'$$
$$A' → ε | \alpha_1 A' | \alpha_2 A' | \ldots | \alpha_n A'$$

Use this rule to eliminate left recursion in the following:
(i)   A → AXYz | z | By
(ii)   A → AXYz | Az | z | By

11. Eliminating Left Recursion. Part III: Exercise 10 describes a rule to eliminate direct left recursion from a general production. To eliminate left recursion from an entire grammar may be more difficult because of indirect left recursion. For example,

```
A → B x y | x
B → C D
C → A | c
D → d
```

is indirectly recursive. (Why?) The following algorithm eliminates left recursion entirely.
   Arrange nonterminals in some order, A_1, A_2, A_3, ... A_n.

```
FOR i := 1 TO n DO
BEGIN {FOR i}
   FOR j := 1 TO n DO
   BEGIN {FOR j}
```

   (1) Replace each production of the form $A_i \rightarrow A_j\gamma$ by the productions:

$A_i \rightarrow \delta_1\gamma | \delta_2\gamma | \ldots | \delta_k\gamma$

   where:

$A_j \rightarrow \delta_1 | \delta_2 | \ldots | \delta_k$

   are all the current A_j productions.
   (2) Eliminate the direct left recursion from the A_i productions:

```
   END {FOR j}
END {FOR i}
```

Eliminate the left recursion from the above grammar.

12. Left Factoring: When a grammar has two or more productions of the form $A \rightarrow \beta\ \alpha_1$, $A \rightarrow \beta\ \alpha_2$, a top-down parser has to look past all the tokens in α in the string being parsed to decide which production to use. Instead the grammar can be changed.   The formal technique is to change

$A \rightarrow \beta\ \alpha_1 | \beta\alpha_2$

to

$A \rightarrow \beta\ A'$
$A' \rightarrow \alpha_1 | \alpha_2$

Apply this technique to:

(a) S → IF C THEN S ELSE S | IF C THEN S
(b) S → REPEAT S UNTIL C | REPEAT S FOREVER

13. Devise a lookahead method (and thus a requirement for grammars to be parsed bottom-up) for bottom-up parsing which will resolve the problem of the handle occurring on the right-hand side of two different productions.

Compiler Project Part II
BNF Grammar
The following is an extended BNF for a subset of Ada:

```
Program                → begin SequenceOfStatements end ;
SequenceOfStatements → Statement {Statement}
Statement              → SimpleStatement
SimpleStatement        → AssignmentStatement
AssignmentStatement  → Name := Expression ;
Name                   → SimpleName
SimpleName             → Identifier
Expression             → Relation
Relation               → SimpleExpression
SimpleExpression       → Term {AddingOperator Term}
Term                   → Factor {MultiplyingOperator Factor}
Factor                 → Primary
Primary                → Name | NumericLiteral | (Expression)
AddingOperator         → + | -
MultiplyingOperator  → * | / | mod | rem
NumericLiteral         → DecimalLiteral
DecimalLiteral         → Integer
```

Assignment
Create a parse tree, by hand, using the grammar above and the program from the project assignment in Chapter 2, repeated below. You need not include every node since the tree is large.

```
begin
    a := b3;
    xyz := a + b + c
     - p / q;
    a := xyz * (p + q);
    p := a - xyz - p;
end;
```

4

Top-Down Parsing

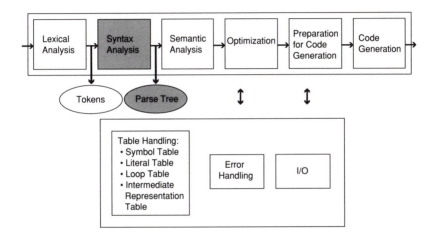

4.0   Introduction

Top-down parsers are often referred to as *predictive* parsers since, at any stage, they try to predict the next lower level of the parse tree. This prediction is done by examining the next token in the input and the current tree and then choosing the production to try next.

Thus the parse tree is built from the top down, trying to construct a left-most derivation as described in Section 3.3.1.

In Chapter 3, grammatical problems for top-down parsers were discussed. The simplest parsing method, recursive descent parsing, cannot be used with grammars that are left recursive.

4.1   Recursive Descent Parsing

Recursive descent parsing uses recursive procedures to model the parse tree to be constructed. For each nonterminal in the grammar, a procedure, which parses a nonterminal, is constructed. Each of these procedures may read input, match terminal symbols or call other procedures to read input and match terminals in the right-hand side of a production.

Consider the expression grammar (with left-recursion removed):

```
E  → T E'
E' → ε | + TE'
T  → FT'
T' → ε | * FT'
F  → (E) | Id
```

In recursive descent parsing, the grammar is thought of as a program with each production implemented as a (recursive) procedure. Thus, in pseudocode, Procedures E and E' would be:

Procedure E

```
BEGIN {E}
  call T
  call E'
  PRINT ("E found")
END {E}
```

Procedure E'

```
BEGIN {E'}
  IF NextSymbol in input = "+" THEN
  BEGIN {IF}
    PRINT ("+ found")
    Advance past it
    Call T
    Call E'
  END {IF}
  PRINT ("E' found")
END {E'}
```

This procedure works fine for correct programs since it presumes that if "+" is not found, then the production is E' → ε.

Procedures T and T' are similar since the grammar productions for them are similar. Procedure F could be written:

Procedure F

```
  BEGIN {F}
  CASE NextSymbol is
"(" :
    PRINT ("( found")
    Advance past it
    Call E
    IF NextSymbol = ")" THEN
    BEGIN {IF}
```

```
              PRINT (") found")
              Advance past it
              PRINT ("F found")
         END {IF}
         ELSE
         error
    "Id" :
         PRINT ("Id found")
         Advance past it
         PRINT ("F found")
    otherwise : error
      END {F}
```

Here, *NextSymbol* is the token produced by the lexical analyzer. To advance past it, a call to get the next token might be made. The PRINT statements will print out the reverse of a left derivation and then the parse tree can be drawn by hand or a procedure can be written to stack up the information and print it out as a real left-most derivation, indenting to make the levels of the "tree" clearer.

4.2 LL Grammars

Recursive descent parsers are a top-down technique similar to LL(1) parsers.

LL(1) parsers use a table to parse, however. The first L stands for the fact that the input string is scanned from left to right. The second L stands for the fact that a left-most derivation is constructed, and the 1 means that one symbol lookahead must be used to create the table. One symbol lookahead just means that the next input symbol is "looked at". LL(1) parsers are essentially table-driven recursive descent parsers.

4.2.1 LL(1) Grammars

To be parsed by LL(1) parsers, a grammar must be LL(1). In order to define LL(1), and to be able to test a grammar to see if it is LL(1), two preliminary definitions are needed: (1) FIRST (α), for strings of terminals and nonterminals α, and (2) FOL-LOW(A) for a nonterminal A.

FIRST(α)

For a string of terminals and nonterminals, α, FIRST(α) is the set of terminals that begins any string derivable from α, including ϵ. Thus, for the standard expression grammar:

```
E → E + T | T
T → T * F | F
F → (E) | Id
```

FIRST(E) = {(, Id} since E can derive strings beginning with "(" or "Id". For "(", E → T → F → (E), a string beginning with "(" derived from "E".

Similarly,

```
FIRST(T * Id) = {(, Id} because T * Id → F * Id → (E) * Id,
                          and T * Id → F * Id → Id * Id
```

The reader can check that

```
FIRST(Id + Id) = {Id}
FIRST(Id) = {Id}
```

Below is the expression grammar with left-recursion removed:

```
E  → T E'
E' → ε  |  + TE'
T  → FT'
T' → ε  |  * FT'
F  → (E)  |  Id
```

Here,

```
FIRST(E) = { (, Id}
FIRST(T * Id) = { (, Id}
FIRST(Id + Id) = {Id}
FIRST(Id) = {Id}
```

Notice that the "firsts" are the same for both versions of the grammar.

FOLLOW(A)

For a nonterminal A in a sentential form, say $\alpha A\beta$, where α and β are some string of terminals and nonterminals,

```
FOLLOW(A) = FIRST(β)
```

That is, FOLLOW(A) is the set of terminals that can appear to the right of A in a sentential form.

For the expression grammar

```
E → E + T | T
T → T * F | F
F → (E) | Id,
FOLLOW(E) = {), +}
```

The ")" is in Follow(E) since "(E)" is a sentential form with ")" following "E". The "+" is in Follow(E) since "E + T" is a sentential form with "+" following the "E".

```
FOLLOW(T) = {+, ), *} :
```

The symbol, "+", is in FOLLOW(T) since E → E + T → T + T. The reader can check that the symbols ")" and "*" are in FOLLOW(T). Using FIRST(α) and FOLLOW(A), LL(1) can now be defined.

LL(1)

Intuitively, a grammar is LL(1) if it is possible to choose the next production by looking at only the next token in the input string. More formally, a grammar is LL(1) if and only if given any two productions A → α, A → β,

(a) FIRST(α) $\cap$ FIRST(β) = $\emptyset$

(b) If one of α or β derives ϵ (clearly both cannot by (a) and the definition of FIRST), say

$\alpha \rightarrow \ldots \rightarrow \epsilon$

then FIRST(β) $\cap$ FOLLOW(A) = $\emptyset$.

EXAMPLE 1   A grammar which is not LL(1)

Consider the following grammar:

```
A → d A
A → d B
A → f
B → g
```

It is not LL(1) because it violates part (a) of the definition: For the two productions, A → d A and A → d B, FIRST(d A) $\cap$ FIRST(d B) = {d} $\neq$ $\emptyset$.

EXAMPLE 2   Another grammar which is not LL(1)

Consider the following grammar:

```
S → X d
X → C
X → B a
C → ε
B → d
```

It is not LL(1) since for the two productions X → C and X → Ba, C → ε and FIRST(Ba) $\cap$ FOLLOW (X) = {d} $\neq$ $\emptyset$

4.3   LL(1) Parsing

4.3.1   The Algorithm

LL(1) parsing is essentially "table driven" recursive descent parsing. Instead of recursive procedure calls, a table is consulted for the next action to perform and an explicit stack is used.

For the moment let us assume that the table is created by magic and consider the following algorithm:

Algorithm
LL(1) Parsing

```
Push "$" onto stack.
Initialize the stack to the start symbol.
REPEAT WHILE stack is nonempty :
  CASE top of the stack is :
```

```
terminal : IF input symbol matches the top of
           the stack, THEN advance input and
           pop stack, ELSE error.
nonterminal : Use nonterminal and current input
              symbol to find correct production
              in table.
              Pop stack.
              Push right-hand side of production
              onto stack, right-most symbol
              first (left-most symbol on top).
END REPEAT
IF input is finished, THEN accept, ELSE error.
```

EXAMPLE 3   LL(1) Parsing

Consider the following grammar, table, and input string:

Grammar:

```
E → T E'
E' → + T E' | ε
T → FT'
T' → * FT' | ε
F → Id | (E)
```

Exercise 7 asks the reader to check that this grammar is LL(1).

The table contains a row for each nonterminal and a column for each terminal.

Table:

INPUT SYMBOLS

		Id	+	*	(	)	$
N O N T E R M I N A L S	E	E → TE'			E → TE'		
	E'		E' → + TE'			E' → ε	E' → ε
	T	T → FT'			T → FT'		
	T'		T' → ε	T' → *FT'		T' → ε	T' → ε
	F	F → Id			F → (E)		

Input: Id * Id + Id

Parsing:

Following the steps in the algorithm, E is first pushed on the stack (the top of the stack is on the right). (See (1).) Since E is on the top of the stack, the nonterminal choice in the CASE statement is taken. The current input symbol is *Id*. In the table, across from E and under *Id* is the production E → TE'. The algorithm says to pop E off the stack and push the right-hand side, TE', onto the stack. This is done in (2). (The top of the stack is on the right.)

In (3), the top of the stack is *Id* and the current input symbol is *Id*. The algorithm says to pop the *Id* off the stack and advance past *Id* in the input. The entire parse is:

Stack	**Input**	**Production**
(1) $E	Id * Id + Id $	E → TE'
(2) $E'T	Id * Id + Id $	T → FT'
$E'T'F	Id * Id + Id $	F → Id
(3) $E'T'Id	Id * Id + Id $	
$E'T'	* Id + Id $	T' → * FT'
$E'T'F*	* Id + Id $	
$E'T'F	Id + Id $	F → Id
$E'T'Id	Id + Id $	
$E'T'	+ Id $	T' → ε
$E'	+ Id $	E' → + TE'
$E'T	+ + Id $	
$E'T	Id $	T → FT'
$E'T'F	Id $	F → Id
$E'T'Id	Id $	
$E'T'	$	T' → ε
$ E'	$	E' → ε
$	$	

Accept

If we construct a derivation using the productions in the Production column, we will construct a left-most derivation.

4.3.2   Constructing LL(1) Parsing Tables

The table is constructed using the following algorithm:

Algorithm
LL(1) Table Construction

For every production A → α in the grammar:

1. If α can derive a string starting with *a*, i.e., for all a in FIRST(α),

    Table[A,a] = A → α

2. If α can derive the empty string, ε, then for all *b* that can follow a string derived from A, i.e., for all *b* in FOLLOW(A),

    Table[A,b] = A → α

Undefined entries are set to error, and if any entry is defined more than once, then the grammar is *not* LL(1).

EXAMPLE 4   Constructing an LL(1) parse table entry using rule 1

Consider the production E → TE' from the non-left-recursive expression grammar of Example 3. FIRST(TE') = {(, *Id*}, so Table[E,(] = E → TE' and Table[E,*Id*] = E → TE' (see table above).

EXAMPLE 5   Constructing an LL(1) parse table entry using rule 2

Consider the production E' → ε from the non-left-recursive expression grammar of Example 3. Since both ")" and "$" are in FOLLOW(E'), this production occurs in the table in the row labeled E' and the columns labeled ")" and "$".

LL(1) parsing tables may be generated automatically. Procedures must be written for FIRST(α) and FOLLOW(A). Then, when the grammar is input, A and α are identified for each production and the two steps above followed to create the table.

4.4   Top-Down Parser Generators

As described in Chapter 3, there are many parser generator tools available today. Although most of these tools generate bottom-up parsers described in the next chapter, there are some top-down tools. It is easier to create a top-down tool than it is to create a bottom-up tool, just as it is easier (for humans) to parse from the top-down than from the bottom-up. One method is to read in the grammar and create an LL(1) parsing table using the method in Section 4.3.2. Then a driver routine can be written to implement the algorithm in Section 4.3.1. Alternatively, a program can be generated instead of a table by implementing the table as a (large) case statement. This is often considered a better approach because it is easier for a user to customize a program than to change a table. If the user is never going to make changes, it is just as easy to generate a table.

4.5   Summary

Top-down parsing is conceptually simple because it is the way we would likely parse by hand. Given an input string and a grammar, a top-down parse expands the start symbol and continues expanding in a left-most derivation until the left-most symbol in the string is generated. Then the next left-most symbol is derived.

Two parsing methods are described in this chapter: recursive descent parsing and LL(1) parsing, a table driven method.

Although top-down parsing is the intuitive way to parse, it is not the most efficient, nor do LL(1) grammars generate the largest class of languages to be parsed. A grammar may need to resolve problems of ambiguity and, in addition, may need to be left factored or have left recursion removed before it can be used in a top-down parsing method. In the next chapter, we will see bottom-up parsing methods, a less intuitive method of parsing, but one that is frequently more efficient. Bottom-up parsing does not place as many restrictions on grammars and can describe a wider range of language constructs than top-down parsing.

EXERCISES

1. Given the following grammar:

   Z → S
   S → **w** S
   S → A B
   A → **x** A
   A → **y**
   B → **z**

   Show a top-down parse (the stack, input and production applied). The first stack and the initial input are shown.

Stack	Input	Production
$Z	wwxyz $	

2. Given the grammar:

   S → CC
   C → aC
   C → b

   and the string

   abab

   Show a top-down parse (the stack, input and production applied). The first stack and the initial input are shown.

Stack	Input	Production
$ S	abab $	

3. Draw the parse tree for the parse in Example 1.

4. The grammar

   S → **x**
   S → **(** S R
   R → **,** S R
   R → **)**

   generates LISP-like expressions such as

   x
   (x,(x,x))
   ((x,(x,x)),x)

   (a) Show that this grammar is LL(1).
   (b) Create an LL(1) parsing table for this grammar.
   (c) Using the table, show a top-down parse (the stack, input and production applied) for (x, (x,x)):

Stack	Input	Production

5. For the grammar:

    ```
    S → xA
    A → y
    S → xC
    C → z
    ```

    (a) Find FIRST(S).
    (b) Is the grammar LL(1)? Why or why not?

6. For the Ada subset grammar at the end of Chapter 3, which is repeated in the project described following these exercises, compute the FIRST and FOL-LOW sets for all the nonterminals.

7. (a) Show that the expression grammar

    ```
    E → E + T | T
    T → T * F | F
    F → (E) | Id
    ```

    is or is not LL(1).

    (b) Show that the expression grammar with left recursion removed

    ```
    E → T E'
    E' → ε | + TE'
    T → FT'
    T' → ε | * FT'
    F → (E) | Id
    ```

    is or is not LL(1).

8. Can there ever be symbols between A and *a* in a sentential form ...A...a... for *a* in FOLLOW(A)?

9. Using the definition of LL(1) as a model, define LL(k), for k > 1.

10. Parsing Non-LL(1) Languages Top-down Anyway: It is sometimes possible to parse a non-LL(1) or even ambiguous language top-down by resolving the choices at parse time. Consider the following grammar which abstracts IF-THEN-ELSE statements (i = IF, t = THEN, e = ELSE, a = terminal statement, c = condition):

    ```
    S → i E t S S'
    S → a
    S' → e S
    S' → ε
    E → c
    ```

    (a) Show that this grammar is not LL(1).
    (b) Create an LL(1) parsing table. (Since the grammar is not LL(1), there will be at least one double entry.)
    (c) Show a top-down parse for
        ```
        IF c THEN IF c THEN a ELSE a
        ```

associating "ELSE" with the closest IF, i.e., choose the table entry that will enforce this:

Stack	Input	Production
$S	IF c THEN IF c THEN a ELSE a $	

Compiler Project Part III
Parsing

Consider the grammar from Chapter 2, repeated here:

```
Program                → begin SequenceOfStatements end ;
SequenceOfStatements   → Statement {Statement}
Statement              → SimpleStatement
SimpleStatement        → AssignmentStatement
AssignmentStatement    → Name := Expression ;
Name                   → SimpleName
SimpleName             → Identifier
Expression             → Relation
Relation               → SimpleExpression
SimpleExpression       → Term {AddingOperator Term}
Term                   → Factor {MultiplyingOperator Factor}
Factor                 → Primary
Primary                → Name | NumericLiteral | (Expression)
AddingOperator         → + | -
MultiplyingOperator    → * | / | mod | rem
NumericLiteral         → DecimalLiteral
DecimalLiteral         → Integer
```

Assignment

Design a recursive descent parser which will implement this grammar.

Run your parser on the following program, repeated from Chapter 2, and on a program of your own design:

```
begin
    a := b3;
    xyz := a + b + c
               - p / q;
    a := xyz * (p + q);
    p := a - xyz - p;
end;
```

Alternate Assignment 1 (Much Easier): Use a parser generator

Using a parser generator of your choice, input the BNF syntax (the tool will describe how to enter it) and create a parser. Use the generated parser to parse the programs described in the regular assignment above.

Alternate Assignment 2 (Harder)

Generate a parser by writing two programs. Follow the description in Section 4.3, using the two algorithms for LL(1) parsing. (1) Write a program that will take the BNF as input, creating an LL(1) table as output, and (2) write a driver routine that will read the table and an input to be parsed, and output a parse tree (or a representation that can be used to build a parse tree). A variant of generating a table is to generate a program that consists of case statements representing the LL(1) table and a driver program that is linked to it to create the LL(1) parser.

5

Bottom-Up Parsing

5.0 Introduction

Bottom-up parsing was introduced in Chapter 3. This chapter describes some bottom-up parsing algorithms.

To review briefly, the parsing problem is to construct the *parse tree* given a *string of symbols* in a language and a *grammar* for that language. Bottom-up parsers build a parse tree from the bottom up, that is, from the leaves up to the root.

5.1 LR Grammars

Corresponding to LL grammars for top-down parsing are LR grammars for bottom-up parsing. The **L** indicates that the input string is scanned from left to right and the **R** indicates that a right derivation is being constructed. Actually, the derivation is constructed in reverse.

5.1.1 LR(k) Grammars

Using the words "handle", "sentential form", and "right derivation" defined in Chapter 3, LR(k) grammars are defined:

A grammar is LR(k) if, for any sentential form in a right derivation,

$$\alpha\beta\omega$$

where ω and α may be null and ω (if non-empty) consists entirely of terminal symbols, the handle β can be identified by looking at the leftmost k symbols of ω.

In practice, $k = 1$.

LR(1) grammars describe constructs typically found in programming languages. Not surprisingly, LR(0) grammars are not often used since LR(0) implies that a string can be parsed without looking at any of the symbols following the handle, and it is difficult to write grammars for the constructs found in programming languages that have this property.

EXAMPLE 1   A grammar which is not LR(0)

The following expression grammar is not LR(0)

```
1. E → E + T
2. E → T
3. T → T * F
4. T → F
5. F → (E)
6. F → Id
```

Consider the sentential form

```
E + T * Id
```

With no lookahead, we would decide that "E + T" is the handle and, using production 1, reduce the sentential form to:

```
E * Id
```

Then *Id* would be the handle which reduces to F and then to T and then to E, leaving us with E * E. This sentential form cannot be reduced. The grammar is LR(1) (see Exercise 3).

EXAMPLE 2   An LR(0) grammar

```
S → a S a
  | b S b
  | c
```

It is harder to show that a grammar *is* LR(k) than that it is not. This example generates symmetric strings of *a*'s and *b*'s with a *c* in the middle. For any sentential form, it is always possible to identify and reduce the handle without looking at any symbols past it.

There is an interesting class of grammars in between LR(0) and LR(1). These are called SLR(1) grammars where "S" stands for simple. They describe many of the constructs found in programming languages. When we create a parser, the only difference among these three (and some others that we will see) is the method used to create the table.

5.2   LR Parsing

LR parsers are efficient (fast) and can find errors as soon as they occur. They are too difficult to build by hand, but there are many tools on the market today that build LR parsing tables given a grammar for the language. We will study LR parsing to get an idea how these tools work.

Like LL parsers, LR parsers may have two parts—a table and a driver.

There are two steps to building LR parsing tables: (1) creating sets of parsing states consisting of item sets and (2) creating the table itself from the states. When no lookahead is used to create the states, and no lookahead is used to create the parse table, then we have an LR(0) parser. One symbol lookahead for both steps is referred to as LR(1).

LR(1) tables can be used to parse typical programming languages, but the tables are extremely large and difficult to construct. There are ways to reduce the size of LR(1) tables, using data structure techniques such as linked lists for sparse matrices. The tables for LR(0) parsers are easy to construct and are small, but describe very little.

The middle ground, SLR(1) parsers, use no lookahead to create the states, but one symbol lookahead to create the table. They describe many of the constructs found in programming languages.

To get a foundation in LR parsing, this chapter will describe SLR(1) parsers. The exercises will explore some other variants. Before describing how to *create* the tables, we will describe how to *use* them (as we did for LL(1) parsing). The parsing algorithm is often described as a *shift-reduce* parsing algorithm.

5.2.1   Shift-Reduce Parsing

LR parsers use a parsing style called *shift-reduce*. This method uses a stack to keep track of what has been parsed already. Based on the stack contents and the input, a decision is made about what action to take next.

As in the section on LL(1) parsing from the previous chapter, let us assume temporarily that parse tables are created by magic.

The table contains two parts; a "shift-reduce" part and a "GO-TO" part. In both cases, rows are labeled by states. The columns in the shift-reduce section of the table are labeled by terminals; the columns in the GO-TO section are labeled by nonterminals.

The table describes four possible actions:

Error: The blank entries in the table. These blank entries indicate a syntax error. No action is defined. (We will expand on this in Chapter 6.)

Accept: Indicated by the "Accept" entry in the table. When we come to this entry in the table, we accept the input string. Parsing is complete.

Shift: Indicated by the "S#" entries in the table where # is a new state. When we come to this entry in the table, we shift the current input symbol followed by the indicated new state onto the stack.

Reduce: Indicated by "R#" where # is the number of a production. The top of the stack contains the right-hand side of a production, the handle. Reduce by the indicated production; consult the GOTO part of the table (across from the current state and under the symbol which occurs on the right-hand side) of production "#" to see the next state, and push the left-hand side of the production onto the stack followed by the new state.

Using these actions, the algorithm is:

Algorithm
LR Parsing

```
Initialize Stack to state 0
Append $ to end of input
While Action ≠ Accept AND Action ≠ Error DO
```

$$\{Stack = s_0x_1s_1.....x_ms_m \text{ and remaining Input} = a_ia_{i+1}.... \$\}$$

{s's are state numbers, x's are sequences of
terminals and nonterminals}

```
    Case Table[s_m,a_i] is
      S#:     Action := Shift
      R#:     Action := Reduce
      Accept: Action := Accept
      Blank:  Action := Error

EndWhile
```

EXAMPLE 3   LR parsing

Consider the following grammar:

```
1. E → E + T
2. E → T
3. T → T * F
4. T → F
5. F → (E)
6. F → Id
```

and table considered built by magic for the moment:

State	Action						GOTO		
	id	+	*	(	)	$	E	T	F
0	S5			S4			1	2	3
1		S6				Accept			
2		R2	S7		R2	R2			
3		R4	R4		R4	R4			
4	S5			S4			8	2	3
5		R6	R6		R6	R6			
6	S5			S4				9	3
7	S5			S4					10
8		S6			S11				
9		R1	S7		R1	R1			
10		R3	R3		R3	R3			
11		R5	R5		R5	R5			

We will use this grammar and table to parse the following input string:

```
id * (id + id)
```

Following are the parsing steps. They are explained following the last step (step 19). The top of the stack is on the right.

	Stack	**Input**	**Action**
(1)	0	id * (id + id) $	S5
(2)	0 id 5	* (id + id) $	R6
(3)	0 F 3	* (id + id) $	R4
(4)	0 T 2	* (id + id) $	S7
	0 T 2 * 7	(id + id) $	S4
	0 T 2 * 7 (4	id + id) $	S5
	0 T 2 * 7 (4 id 5	+ id) $	R6
	0 T 2 * 7 (4 F 3	+ id) $	R4
	0 T 2 * 7 (4 T 2	+ id) $	R2
	0 T 2 * 7 (4 E 8	+ id) $	S6
	0 T 2 * 7 (4 E 8 + 6	id $	S5
	0 T 2 * 7 (4 E 8 + 6 id 5	) $	R6
	0 T 2 * 7 (4 E 8 + 6 F 3	) $	R4
	0 T 2 * 7 (4 E 8 + 6 T 9	) $	R1
	0 T 2 * 7 (4 E 8	) $	S11
	0 T 2 * 7 (4 E 8) 11	$	R5
	0 T 2 * 7 F 10	$	R3
	0 T 2	$	R2
(19)	0 E 1	$	Accept

Step (1)

Parsing begins with state 0 on the stack and the input terminated by "$":

	Stack	**Input**	**Action**
(1)	0	id * (id + id) $	

Consulting the table, across from state 0 and under input **id** is the action **S5** which means to **s**hift (push) the input onto the stack and go to state 5.

Step (2)

Stack	Input	Action
(1) 0	id * (id + id) $	S5
(2) 0 id 5	* (id + id) $	

Consulting the table, across from state **5** and under input **∗**, we find the action **R6** which means the right-hand side of production 6 is the handle to be reduced. We remove everything on the stack that includes the handle. Here, this is **id 5**. The stack now contains only **0**. Since the left-hand side of production 6 will be pushed on the stack, consult the GOTO part of the table across from state **0** (the exposed top state) and under **F** (the left-hand side of the production). The entry there is **3**. Thus, we push **F 3** onto the stack.

Step (3)

Stack	Input	Action
(1) 0	id * (id + id) $	S5
(2) 0 id 5	* (id + id) $	R6
(3) 0 F 3	* (id + id) $	

Now the top of the stack is state **3**, and the current input is **∗**. Consulting the table, across from **3** and under **∗**, the action indicated is R4—reduce using production 4. Thus, the right-hand side of production 4 is the handle on the stack. The algorithm says to pop the stack up to and including the F. That exposes state 0. Across from 0 and under the right-hand side of production 4 (the T) is state 2. We shift the T onto the stack followed by state 2.

We encourage the reader to continue with the remaining steps. We will show the last step, step 19:

Step (19)

The parse is in state 1 looking at "$". The table indicates that this is the accept state. Parsing has thus been completed successfully. By following the reduce actions in reverse, starting with **R2**, the last reduce action, and continuing until **R6**, the first reduce action, a parse tree can be created.

At any stage of the parse, we will have the following configuration:

Stack	Input
$s_0 x_1 s_1 \ldots \ldots x_m s_m$	$a_i a_{i+1} \ldots \ldots$ $

where the s's are states, the x's are sequences of terminals and nonterminals, and the a's are input symbols. This is somewhat like a finite-state machine where the state on top (here, the right) of the stack contains the "accumulation of information" about the parse until this point.

We just have to look at the top of the stack and the symbol coming in to know what to do.

We can construct such a finite state machine from the productions in the grammar where each state is a set of items.

5.2.2   Creating the Table

Items

An item is a production with a *position marker*, (•), e.g.,

E → E • +T

which indicates the state of the parse where we have seen a string derivable from E and are looking for a string derivable from +T.

Items are grouped and each group becomes a state which represents a condition of the parse. We will state the algorithm and then show how it can be applied to this example.

Algorithm
Constructing States

(0) Create a new nonterminal S' and a new production S' → S where S is the *start* symbol

(1) **IF** S is the *start* symbol, put S' → • S into a *start state* called *state 0*.

(2) Closure: **IF** A → x • Xα is in the state, **THEN** add X → • ω to the state for every production X → ω in the grammar.

(3) Creating a *new* state from an *old* state: Look for an item of form A → x•zω where z is a single terminal or nonterminal and build a new state from A → xz•ω. (Include in the *new* state all items with •z in the *old* state.) A new state is created for each different z.

(4) Repeat steps 2 and 3 until no *new* states are created. (A state is *new* if it is not identical to an old state.)

EXAMPLE 4   Constructing states for the expression grammar

Step 0

Create E' → E

Step 1

State 0
E' → • E

Step 2

E' → • E fits the model A → x • X ω, with x, ω = ε, and X = E. E → T and E → E + T are E productions; thus E → • T and E → • E + T are added to state 0.

State 0
E' → • E
E → • E + T
E → • T

Reapplying step 2 to E → • T adds

T → • T * F
T → • F

and reapplying step 2 to T → • F adds

F → • (E)
F → • Id

State 0 is thus:

State 0
E' → • E
E → • E + T
E → • T
T → • T * F
T → • F
F → • (E)
F → • Id

If the dot is interpreted as separating the part of the string that has been parsed from the part yet to be parsed, state 0 indicates the state where we "expect" to see an E (an expression). Expecting to see an E is the same as expecting to see an E + T, i.e., the sum of two things, or a T (a term), or a T * F, i.e., the product of two things, etc. since all of these are possibilities for an E (i.e., an expression).

Using step 3, there are two items in state 0 with an E after the dot. E' →• E fits the model A → x•zω with x and ω = ε, z = E. Thus, we build a new state, putting E' —.> E• into it. Since E → • E + T also has E after •, we add E → E • + T. Step 2 doesn't apply, and so we are done with state 1.

State 1
E' -> E •
E -> E • + T

Interpreting the dot, •, as above, the first item here indicates that the entire expression has been parsed. When we create the table, this item will be used to create the "accept" entry. (In fact, looking at the table above, it can be seen that "accept" is an entry for state 1.) Similarly, the item E → E • + T indicates the state of a parse where an expression has been seen and we expect a "+ T". The string might be "Id + Id" where the first *Id* has been read, for example, or *Id * Id + Id * Id* where the first *Id * Id* has been read (see Exercise 8).

Continuing, the following states are created (see Exercise 9):

State 2	State 3	State 4	State 5	State 6
E -> T •	T -> F •	F -> (• E)	F -> Id •	E -> E + • T
T -> T • * F		E -> • E + T		T -> • T * F
		E -> • T		T -> • F
		T -> • T * F		F -> • (E)
		T -> • F		F -> • Id
		F -> • (E)		
		F -> • Id		

State 7	State 8	State 9	State 10	State 11
T -> T * • F	F -> (E •)	E -> E + T •	T -> T * F •	F -> (E) •
F -> • (E)	E -> E • + T	T -> T • * F		
F -> • Id				

These are called *LR(0) items* because no lookahead was considered when creating them.

We could use these to build an LR(0) parsing table, but for this grammar, there will be multiply defined entries since the grammar is not LR(0) (see Example 1). These can also be considered SLR items and we will use them to build an SLR(1) table, using one-symbol lookahead.

Algorithm
Construction of an SLR(1) Parsing Table

(1) Shift: IF A → x • a ω is in state *m* for input symbol a, AND A → x a • ω is in state *n*, THEN enter S*n* at Table[*m,a*].
(2) Reduce: IF A → ω • is in state *n*, THEN enter r*i* at Table[*n,a*] WHERE *i* is the production *i*: A → ω and a is in FOLLOW(A).
(3) Accept: IF S' → S • is in state *n*, THEN enter "accept" at Table[*n*,$].
(4) GOTO's: IF A → x • B ω is in state *m*, AND A → x B • ω is in state *n*, then enter *n* at Table[*m,B*].

EXAMPLE 5   Creating an SLR(1) table for the expression grammar

Following are some of the steps for creating the SLR(1) table shown in Example 4. One example is shown for each of the steps in the algorithm.

Step 1

E → E • + T is in state 1
E → E + • T is in state 6
so Table[1,+] = S6.

Step 2

In state 3 we have T → F •

The terminals that can *follow* T in any sentential form are +, *,), $ (see Chapter 4). So Table[3,+] = Table [3,*] = Table[3,)] = Table[3,$] = R4 where 4 is the number of production T → F.

Step 3

E' is the *start* symbol here and E' → E • is in state 1, so Table[1,$] = "accept".

Step 4

E → • E + T is in state 0, while E → E • + T is in state 1, so Table[0,E] = 1.

5.2.3   Shift-Reduce Conflicts

Sometimes the parser does not know whether to continue with a production rule or to accept an entirely different rule.

A shift-reduce conflict occurs when the parser cannot decide whether to continue shifting symbols onto the stack or to reduce a handle which is already on the stack; that is, if the grammar is not SLR(1) (or LR(1) for an LR(1) table), then more than one entry may occur in the table. If both a "shift" action and a "reduce" action occur in the same entry, and the parsing process consults that entry, then a *shift-reduce* conflict is said to occur. One way to resolve such conflicts is to attempt to rewrite the grammar. The second method is to analyze the situation and decide, if possible, which action is the correct one. If neither of these steps solves the problem, then it is possible that the underlying language construct cannot be described using an SLR(1) (or LR(1)) grammar, and another method will have to be used.

5.2.4   Reduce-Reduce Conflicts

Sometimes the parser does not know which of several equally acceptable rules to use.

A reduce-reduce conflict occurs when the handle on the stack occurs on the right of more than one production. As with shift-reduce errors, if the grammar is not SLR(1) (or LR(1) for an LR(1) table), then more than one entry may occur in the table. If two different "reduce" actions occur in the same entry, and the parsing process consults that entry, then a *reduce-reduce* conflict is said to occur. These conflicts may also be resolved by attempting to rewrite the grammar or by analyzing the particular situation; if neither is possible, then the underlying language is not SLR(1) (or LR(1)).

5.2.5   LR Parser Variants

In Section 5.2.2, LR(0) item sets were created; that is, no lookahead was used to create them. We did, however, consider the next input symbol (one symbol lookahead) when creating the table (see Algorithm SLR). If no lookahead is used to create the table, then the parser would be called an LR(0) parser. Unfortunately, LR(0) parsers don't recognize the constructs one finds in most programming languages. If we consider the next possible symbol for each of the items in a state, as well as for creating the table, we would have an LR(1) parser.

LR(1) tables for typical programming languages are massive. SLR(1) parsers do recognize most, but not all, of the constructs in typical programming languages.

There is another type of parser which recognizes almost as many constructs as an LR(1) parser. This is called a LALR(1) parser and is constructed by first constructing the LR(1) states and then merging many of them. Whenever two states are the same except for the lookahead symbol, they are merged (see Exercises 19 and 20). The first *LA* stands for *lookahead* since a lookahead token is added to the item.

5.2.6   LR Parser Generators

By now, it should be quite clear that creating LR parsers "by hand" for real programming languages is too difficult. That is the bad news. The good news, however, is that it is not necessary. There are many LR parser generators available. Perhaps the most famous is *YACC*, the LALR(1) tool that comes with the UNIX operating system. *YACC* stands for *yet another compiler compiler* which should reinforce the notion that there are a lot of such tools.

Another well-known parser generator is *TWS, translator writing system,* which is LR(1).

QParser is a LALR(1) parser generator that can be purchased for the IBM PC.

Once again, the only difference between any of these is in the way the tables are built:

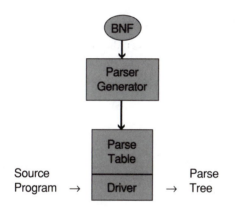

(Many of the tools generate the table in program form so that changes can be made by the user.)

The reader might question why there are so many such tools. A survey and study (Lemone, 1987) revealed that parser generators are often quite difficult to use, even for computer science professionals, and that the input language (called a metalanguage) for expressing the BNF is often difficult to learn. A second problem is that such tools are not well documented, making them still more difficult to use. Yet a third problem is the lack of error analyses when errors occur. It appears that even though there are a lot of such tools on the market, a substantial improvement can still be made in their design.

5.3   Operator Grammars

Operator grammars describe special purpose languages or subsets of typical programming languages. For example, expressions can be described using operator grammars (thus we can continue using the same example . . .). The programming language SNOBOL can be described completely using an operator grammar.

Definition of an Operator Grammar:
(1) Has no ε-rules, that is, no rules of the form A → ε
(2) Has no adjacent nonterminals, that is, no rules of the form A →...XY...

EXAMPLE 6   A non-operator grammar

```
E → EAE | Id
A → + | -
```

This grammar is *not* an operator grammar since the first production has three adjacent nonterminals.

This grammar can be rewritten as:

```
E → E + E | E - E| Id
```

which *is* an operator precedence grammar, although it is also ambiguous.

5.3.1   Operator Precedence Parsing

Precedence relations can be set up between the operators in an operator grammar. In operator precedence, the string is scanned from left to right as usual and there are essentially two states:

> *State 1:* "expecting an operator"

or

> *State 2:* "expecting an operand"

To parse, we just have to keep two stacks—one for operators and one for operands—and compare the precedence of consecutive operators. When an operator of higher precedence (than the next operator down in the stack) is on top, then the handle consists of the two operands combined with that operator.

EXAMPLE 7   Simple precedence parsing

Consider the string:

```
a + b * c $
```

and the grammar:

```
1. E → E + E
2. E → E * E
3. E → (E)
4. E → Id
```

Since (we know that) * takes precedence over +, and we will presume that "$" has the lowest precedence, we don't need the extra symbols T and F because they are there to establish the precedence of the operators. The grammar is ambiguous, but the method will give only one parse tree:

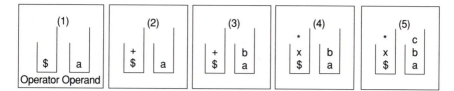

The boxes labeled (1) to (5) show operators and operands being pushed onto their respective stacks. Since the top of the stack in (5) contains an operator (*) of higher precedence than the next lower element (+), the handle must be $b * c$. Exercise 15 continues this example.

In this section, precedence is applied between the operators, and no table has been consulted. This is fine for a limited number of operators, but for a larger number, a table is more efficient.

5.3.2 Table Driven Operator Precedence Parsing

Precedence Relations

We formalize precedence relationships by introducing three disjoint precedence relations:

$$<\bullet \quad (\doteq) \quad \bullet>$$

where $a <\bullet b$ means a has lower precedence than b, $a \doteq b$ means a has the same precedence as b, and $a \bullet> b$ means a has higher precedence than b. Thus for expressions, $+ <\bullet *$, $(\doteq)$, and $* \bullet> +$.

As in Section 5.3.1, we identify the handle by finding two symbols, a, b, such that $a \bullet> b$.

The algorithm is simple:

> Algorithm
> *Precedence Parsing*
>
> ```
> Insert precedence relations
> While Input ≠ $S$ DO
>    Scan to leftmost •>.
>    Scan back to first <• to the left.
>    Reduce the handle that is between.
>    Reinsert precedence relations ignoring nonterminals
> EndWhile
> ```

EXAMPLE 8 Operator precedence parsing

Consider the grammar:

$$E \rightarrow E + E$$
$$E \rightarrow E * E$$
$$E \rightarrow (E)$$
$$E \rightarrow a$$

and input string:

    (a + a)

It is usual to surround the input with $, and $ has lowest precedence, i.e., $<• +, etc.

As usual, the table will be built magically for us to use (and then we will discuss how to build one):

Table:

	(	a	*	+	)	$
)			•>	•>	•>	•>
a			•>	•>	•>	•>
*	<•	<•	•>	•>	•>	•>
+	<•	<•	<•	•>	•>	•>
(	<•	<•	<•	<•	≐	
$	<•	<•	<•	<•		≐

We use the table to insert the precedence relations between the elements of the string:

    $ <• (<• a •> + <• a •>) •> $

Scanning to the first **•>**, and then back to the nearest **<•** :

    $ <• (<• a •> + <• a •>) •> $
              | |

identifies (the first) a as the handle. Using the grammar, we reduce this to E:

          E
          |
    $ (a + a) $

The sentential form is now:

    $ (E + a) $

Ignoring nonterminals for reinserting the precedence relations gives

    $ <• (<• + <• a •>) •> $
               | |

and (the second) a is now the handle:

          E   E
          | |
    $ (a + a) $

and the sentential form is:

    $ (E + E) $

Since the algorithm says to continue until the sentential form is E (E is the start symbol here), we reinsert the precedence relations, again ignoring the nonterminals:

$$\$ \ <\bullet \ (\ <\bullet \ + \ \bullet> \ ) \ \bullet> \ \$$$

Thus the *handle* lies between *(* and *)*; that is "E + E" is the handle, and the parse becomes:

and the sentential form is:

$$\$ \ (E) \ \$$$

Reinserting the precedence relations,

$$\$ \ <\bullet \ (\ \dot= \ ) \ \bullet> \ \$$$

identifies "(E)" as the handle.
The parse tree is:

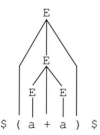

The sentential form is:

$$\$ \ E \ \$$$

and we are finished.

5.3.3   Building the Table

The method presented here exploits the following fact:

Operators lower in the parse tree have higher precedence.

The reader is encouraged to study the following parse tree and to build a few of her own, using the expression grammar to become convinced of this. For clarity, we will use the standard expression grammar with T's and F's:

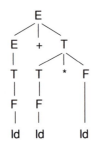

We use the observation above to make the following definitions. These definitions will then be used to build the table.

Definition 1: *Leading* (N), where N is a nonterminal, is the set of all terminals that can appear first in a sentential form derivable from N.

Definition 2: *Trailing* (N), where N is a nonterminal, is the set of all terminals that can appear last in a sentential form derivable from N.

EXAMPLE 9   Computing leading and trailing sets for the expression
                 grammar

Consider, once again, the expression grammar:

```
1. E → E + T
2. E → T
3. T → T * F
4. T → F
5. F → (E)
6. F → Id
```

Then Leading(E) = {+, *, (, *Id* }

+   is in Leading(E) because E + T is a sentential form derivable
     (directly) from E.

*   is in Leading(E) because T * F is a sentential form derivable from
     E (E→T→T * F), and * is the first (and only) terminal.

(is in Leading(E) because (E) is a sentential form derivable from E
     (E→T→F → (E)) and (is the first terminal.

Id   is in Leading(E) because *Id* is a sentential form derivable from E
     (E→T→F→*Id*) and *Id* is the first (and only) terminal.

Similarly,

+   is in Trailing(E) because it is the *only* terminal in the sentential
     form E + T and thus the *last* terminal in E + T (as well as the *first*).

)   is in Trailing(E) because (E) is a sentential form derivable from E
     (E→T→F→(E)) and) is the last terminal in this sentential form.

Exercise 16 asks for the rest of the leading and trailing sets for this grammar.

The precedence table can be built by noting that for an operator grammar G and a production of the form

$$N \rightarrow \alpha \; A \; t \; B \; \beta$$

where t is a terminal and A and B are nonterminals,

t always appears higher in a parse tree than Leading(B)

This is true because deriving a string from B will place the string, and hence B's leading set, lower in the tree.

Similarly,

Trailing(A) always appears lower in a parse tree than t

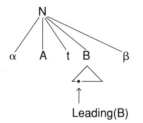

Leading(B)

This is true because deriving a string from A will place the string and hence A's trailing set lower in the tree.

Thus,

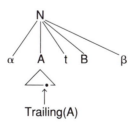

Trailing(A)

Rule 1 $t <\!\bullet\; Leading(B)$

Rule 2 $Trailing(A) \;\bullet\!> t$

Thus, to compute the precedence relations for a grammar G, we consider each production in turn, identify t, A, and B and apply (1) and (2). Note that if we come up with more than one precedence relation between a pair of terminals, then we do not have an operator precedence grammar.

One more rule is needed to compute the table:

Rule 3   If terminals t_1 and t_2 both appear on the right-hand side of the same production, then

$$t_1 \doteq t_2$$

For our example "(" and ")" both appear in F → (E). Thus

$$(\doteq)$$

The algorithm is now simple:

Algorithm
Creating a Precedence Table

```
FOR each production of the form N → α A t B β,
   Compute Leading(B)
   Compute Trailing(A)
   Use these sets to compute precedence using Rule 1,
   Rule 2 and Rule 3. "$" <• all other terminals.
ENDFOR
```

Example 10   Computing precedence relations for the expression grammar

We consider the two productions E → E + T and T → T * F and compute leading and trailing of E, T and F:

NT	Leading	Trailing
E	+, *, (, Id	+, *,), Id
T	*, (, Id	*,), Id
F	(, Id	), Id

Grammar:
E → E + T | T
T → T * F | F
F → (E) | ID

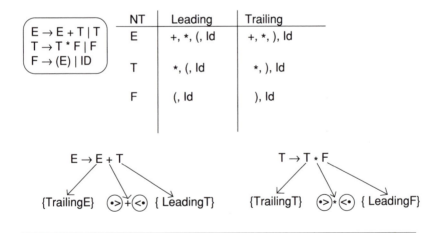

There is only one more production to consider in creating the table because the productions with just one terminal or nonterminal on the right-hand side are not involved in creating the precedence relations (see Exercise 16).

Most grammars are *not* operator precedence, but operator grammars can sometimes be written for small special purpose applications. Also, an easy way to improve an LL parser for a programming language is to parse the expressions using operator precedence.

5.4   Choosing a Parsing Method

The last three chapters have discussed numerous issues related to parsing and presented a variety of algorithms. Any of the methods might be chosen, but some are better under some circumstances.

5.4.1   Choose a Parser Generator

Everything else being equal, the best method may be to use a parser generator tool. Since there are a wide variety of these, a further decision might still need to be made. Certainly, the tool should run on the user's machine, and many, but not all, tools come in versions for different machines.

An LR(1) tool can handle a wider variety of languages than a LALR(1) tool, but the tables are sometimes monstrous. An LL(1) tool can generate parsers for almost as many languages as LALR(1), and LL(1) tools are, as we have seen, easier to understand and thus the resulting parser (or table) will be easier to modify, if necessary. The major drawback to LL(1) is that left-recursion, if any, must be removed from the grammar, and the grammar may need to be left factored.

5.4.2   Build Your Own Parser

If the parser needs to be built fast, then designing and implementing a recursive descent parser is probably the best bet. It is the simplest method and semantic decisions can be easily incorporated, as we shall see in Chapter 7. It is no coincidence that many classroom compiler projects assign a recursive descent parser to be built. On the other hand, since the parser is in the code (no table), it is very difficult to make changes. In addition, as we will see in Chapter 6, error handling is somewhat difficult. Again, left-recursion elimination and left-factoring must be done, if necessary, before the grammar can be used.

Another alternative is to build an LL(1) parser or even an LL(1) parser generator.

For both recursive descent and LL(1), the efficiency of the parser can be improved by parsing the expressions using the operator precedence method.

No one should ever consider writing an LR parser except for perhaps a small language that can be parsed using an SLR(1) parser. Even designing and building an LR parser generator is a project to be undertaken only if it is to be used and reused.

Operator precedence parsers are ideal for (the rare, but not nonexistent) languages that can be described using an operator grammar.

5.5   Summary

This chapter has been a discussion of bottom-up parsing techniques, also called shift-reduce parsing techniques. Two methods, LR and operator precedence, are discussed.

LR parsers come in two parts: a table and a driver, and three flavors: simple LR (SLR), LR (sometimes termed canonical LR), and lookahead LR (LALR). Only SLR is discussed in detail in the chapter, the others being described in the exercises. LR(1) parsers parse the constructs typically found in programming languages, but may contain thousands of states. LALR(1) parsers still parse most of the constructs, but contain fewer states.

LR parser tools are plentiful, but vary in their ease of use, documentation, and functionality.

Operator precedence parsers parse expressions and some special purpose languages that can be described using operator grammars.

EXERCISES

1. Show that the following grammar is not LR(0):

    ```
    List → Id [ELList]
    ELList → Id
    ELList → ELList , Id
    ```

2. Show that the following grammar is LR(0):

    ```
    S → x
    S → ( S R
    R → , S R
    R → )
    ```

3. Show that the expression grammar (see Example 3) is LR(1).

4. Show that the expression grammar (see Example 3) is SLR(1).

5. Devise a grammar which is LR(1) but not SLR(1).

6. Draw the parse tree for the parse described in Example 1.

7. Why is a new nonterminal, S', created, and a new production, S' → S, added to the list of productions in creating LR parsing tables? (See the algorithm for creating states.)

8. Consider an item of the form E → E • + T. List five strings to which this item might apply and indicate which part of the string has been read and which part is still expected to be read.

9. Verify, by creating them, the states for Example 4.

10. How is lookahead used for
    (a) SLR(1)? (Hint: Look at the SLR(1) algorithm.)
    (b) LR(1) ?

11. Consider the grammar:

    ```
    1) S → X X
    2) X → x X
    3) X → y
    ```

    (a) Create the LR(0) items and an SLR(1) parsing table.
    (b) Using the table created in (a), parse the string y x x y.

Stack	Input	Action
0	y x x y $	

12. Consider the following grammar:

```
S → x
S → ( S R
R → , S R
R → )
```

which generates LISP-like expressions such as

```
    x
(x, (x, x))
((x, (x, x), (x, x), x)
```

(a) Augment the grammar with S' → S and calculate the LR(0) items and an LR parsing table.

(b) Parse

```
(x, x)
```

13. Parsing Ambiguous Grammars: Consider the following (ambiguous) grammar which abstracts IF THEN ELSE statements:

```
1. S → i E t S S'
2. S → a
3. S'→ e S
4. S'→ ε
5. E → b
```

where:

```
i: IF
t: THEN
e: ELSE
a: statement (terminal)
b: condition (terminal)
```

(a) Augment the grammar with S" → S, and calculate the LR(0) items.

(b) Create an SLR(1) parsing table. Is the grammar SLR(1)? Why or why not?

(c) Parse the string "IF b THEN IF b THEN a ELSE a" resolving the ambiguity by associating ELSE with the closest IF.

14. From the method described in Example 8, (a) write an algorithm for performing simple precedence parsing, and (b) use your algorithm to parse the string in the example.

15. Compute the leading and trailing sets for the expression grammar, corroborating the results in Example 10.

16. Considering the one appropriate remaining production not considered in Example 9, compute the precedence relations.

17. Create the precedence table for the following grammar:

```
S → T | S & T
T → U | T # U
U → Id
```

18. LR(1) Parsing: An LR(1) item is an LR(0) item plus a set of lookahead characters, e.g.,

```
E → E • + T, {$,+}
```

which indicates that we have seen an E and are expecting a "+ T", which may then be followed by the end of string (indicated by $) or by a "+" (as in $a + a + a$).

The algorithm is the same as for creating LR(0) items except the closure step which now needs to be modified to include the lookahead characters:

```
(2) (closure) IF A → x • X y, L is in the state
    THEN add X → • z, FIRST(yl) for each l in L to
    the state for every X → z.
```

We build the first two states here and leave the remaining (21) to the reader.

State 0:   E' → • E, {$}. {$} indicates that the string is followed by $. Applying the closure rule to this gives us initially E → • E + T , {$} as the next item here since FIRST (ϵ $) = {$}. Now, the closure operation must be applied to this and FIRST (+ T $) = {+}, so the next item is E → • E + T, {+, $}. The entire states 0 and 1 are:

State 0

```
E' → • E, {$}
E  → • E + T, {+, $}
E  → • T, {+, $}
T  → • T * F, {$, +, *}
T  → • F, {$, +, *}
F  → • Id, {$, +, *}
F  → • (E), {$, +, *}
```

State 1

```
E' → E • , {$}
E  → E • + T, {+, $}
```

(a) Compute the rest of the states for this grammar.

(b) Create the LR(1) parsing table. Note that for items in state i of the form A → ω •, {l_1, l_2 ...}, Table[i, l_1] = Table[i, l_2] = ... = R# where # is the number of the production A → ω.

(c) Use the table to parse the string *Id * (Id * Id)*.

19. LALR(1) Parsing: It is often the case that two states in an LR(1) state set have the exact same items except for the lookahead. We can reduce the size of the ultimate table by merging these two states. There are ten pairs of states that can be merged in Exercise 18. Two of them and their merged state are:

State i

```
E → E + • T, {+, $}
T → • T * F, {$, +, *}
T → • T * F, {$, +, *}
F → • Id, {$, +, *}
F → • (E), {$, +, *}
```

State j

```
E → E + • T, {), +}
T → • T * F, {), +, *}
T → • T * F, {), +, *}
F → • Id, {), +, *}
F → • (E), {), +, *}
```

State i-j

```
E → E + • T, {), +, $}
T → • T * F, {), +, *, $}
T → • T * F, {), +, *, $}
F → • Id, {), +, *, $}
F → • (E), {), +, *, $}
```

(a) Create the LALR(1) items sets for the expression grammar from the state sets in Exercise 18.

(b) Create the table from the items. Note the difference in size (since there are fewer states).

(c) Parse the string *Id* ∗ *(Id + Id)*.

Compiler Project Part IV

Enhancement

Enhance your compiler to scan and parse programs described by the grammar of Part III (see Chapter 4) with the addition of the following productions:

```
Statement            → SimpleStatement | CompoundStatement
Compound Statement   → IfStatement | LoopStatement
If Statement         → if Condition then SequenceOfStatements
                         {elsif Condition then SequenceOfState-
                            ments}
                         [else SequenceOfStatements]
                         end if ;
Loop Statement       → [IterationScheme] loop SequenceOfState-
                            ments
                         end loop ;
Iteration Scheme     → while Condition
Condition            → Expression
Expression           → Relation {and Relation} |
                         Relation {or Relation} |
                         Relation {xor Relation}
Relation             → SimpleExpression
                         [RelationalOperator SimpleExpression]
Relational operator  → = | /= | < | <= | > | >=
```

Run your compiler on the following input:

(a)
```
begin
   if i > j then
      i := i + j;
   elsif i < j then
      i := 1;
   end if;
end;
```

(b)
```
begin
   while (i < j) and (j < k) loop
      k := k + 1;
      while (i = j) loop
         i := i + 2;
      end loop;
   end loop;
end;
```

(c) A program of your choice.

6

Error Handling

"There are more incorrect programs than correct ones"

—Unknown

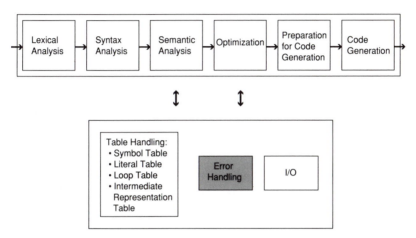

6.0 Introduction

Most compilers spend a large amount of time and space handling errors. Half the code space of many compilers is devoted to error handling. Error detection, reporting and recovery are noticed by the average user far more than the speed of a compiler or the speed of the emitted code.

There are actually four facets to error handling:

- Error Creation
- Error Detection
- Error Reporting
- Error Recovery

6.1 Error Creation

The design of a language affects the kind of errors that occur. Example 1 shows the FORTRAN Hollerith statement, an early (but still legal today!) way of specifying a character string to be output.

EXAMPLE 1   FORTRAN Hollerith

39HExactly 39 characters must be in string

In Example 1, the number preceding the H had to be an exact count of the number of characters following the H. This is very conducive to error creation as many FOR-TRAN programmers will attest.

Consider Example 2 which shows quoted strings which are allowed to go over line boundaries as in the programming language Pascal.

EXAMPLE 2   Quoted strings

"If no closing quote, then the whole rest of the program is part of the character string

```
I := 3
.
.
.
```

Here, the programmer didn't terminate the string when she went on to a new line. Again this is a very easy error to make. A better language design, one followed by newer languages such as Ada, is not to allow quoted strings to run over line boundaries.

Other language constructs can encourage errors *not* picked up by the compiler. Example 3 shows another easy way to make errors. These were probably made by a programmer adding comments after the program is written (better late than never!).

EXAMPLE 3   Comments

What the programmer meant to code:

```
I := 3; {Comment}
J := 2; {Comment}
K := 1; {Comment}
```

What was actually coded:

```
I := 3; {Comment
J := 2; Comment
K := 1; Comment}
```

In a language that allows comments to run over line boundaries (such as Pascal), this is syntactically correct. Unfortunately, only the first assignment statement will ever be executed.

PL/I is a very flexible language in which it is easy to make errors not detected by the compiler. In Example 4, the programmer coded an inequality much as it is written in mathematics.

EXAMPLE 4   Inequalities in PL/I

What the programmer meant:

```
IF 0 < I AND I < 10 THEN ...
```

What was coded:

```
IF 0 < I < 10 THEN ...
```

Unfortunately, this last inequality is always true in PL/I: IF 0 < I is true, PL/I assigns a value of 1; if 0 < I is false, PL/I assigns a value of 0. Either of these is less than 10, so the inequality is always true!

EXAMPLE 5   Equality check in C

What the programmer meant:

```
if (x == 0) {...}
```

What was coded:

```
if (x = 0) {...}
```

The double equal sign, when testing for equality, is often a source of C syntax errors.

6.2 Error Reporting

The first compilers weren't very tactful when it came to error reporting. It is reputed that one of the early FORTRAN compilers issued the message

```
"ERROR"
```

upon encountering one or more syntax errors. Similarly, the Joss (another early language) compiler used to report:

```
"EH?"
```

In order to be of use, an error message should inform the user *where* the error is and, as much as possible, *what* the error is. More formally, error messages should report the *source* of the error and *parametrize* the message.

For example, the message

```
In line 2, reserved word initiate is misspelled
    |                 \   /
  Source          Parameters
```

Here, the source is "line 2", and the parameters are "reserved word" and "initiate". An error handler can be written in a modular way when it contains a small number of schema such as

In *Source Type Value* is missing

When the handler is called, actual values are substituted for the templates *Source*, *Type*, and *Value*.

Another error reporting issue is whether to continue reporting repeated instances of the same error. Should an error message be issued every time an undeclared variable is used? A poll of the author's class revealed an even split "pro and con" on this issue. It is clearly a matter of taste.

6.3 Error Detection and Recovery

Error detection and recovery are a combined issue in that the detection of an error often gives a strong clue as to how to recover from it. Error detection finds the error; error recovery tries to mend the error enough to continue the parse (and perhaps find more errors). Recovery is somewhat heuristic in that it is very difficult to be sure that the "correction" does not report spurious errors (ones that weren't there) or ignore other errors that were there.

Fortunately, for most compilers, the most common errors are also the simplest. Missing reserved words such as BEGIN and END and missing delimiters such as semicolons are common, as are misspellings. (A very common error, undeclared variable names, will be discussed in Chapter 7.)

A simple error recovery is to skip to the end of the current construct or to the beginning of the next. This is called *panic mode* recovery. Here, the error handler peruses the input following the error looking for a symbol that ends a construct such as ";" or "END". Alternatively, the input can be perused for a symbol that starts something such as "BEGIN" or "IF".

Of course, other errors may be skipped in the process.

A more sophisticated method is to replace a prefix contained in the remaining input by some string that allows the parser to continue. This is done not to try to correct the program, but so that the parser can continue, perhaps finding more errors.

LL and LR parsers have the *viable-prefix property*, that is, they detect that an error has occurred as soon as they see a prefix of the input that is not a prefix of any string in the language. Recovery frequently involves adding to or removing from the input. Alternatively, one could add to or remove from the parsing stack.

6.3.1 LR Error Detection and Recovery

Consider the following grammar from Exercise 12 of Chapter 5:

```
1) S → X X
2) X → x X
3) X → y
```

The parse table is

Parse Table

State			Action		GOTO	
	x	**y**	**$**		S	X
0	s3	s4			1	2
1			accept			
2	s3	s4	err3		5	
3	s3	s4			6	
4	r3	r3	r3			
5	r1					
6	r2	r2	r2			

In State 2 under "$" is the entry "err3", that is, if the parse is in State 2 at the end of the input, that is an error.

Suppose the string to be parsed is *x y*. The following is the state of the parse when it discovers that what follows (the end of string marker) is not what is expected:

STACK	INPUT
0x3y4	$
0x3X6	$
0X2	$

It is easy to see that err3 might be:

```
X expected
```

Recovery might consist of adding a string derivable from X, i.e., a *y* to the input or, equivalently, pushing *y* onto the stack, followed by state 3.

6.3.2   LL Error Detection and Recovery

LL parsing is similar to bottom-up in that we have input and a stack.

Consider the following grammar and input:

Grammar	Input
S → aAS	ab...
S → bA	
A → cA	
A → d	

A top-down parse would proceed (with the stack "top" on the right):

Stack	Input
$ S	ab...
$ SAa	ab...
$ SA	b...

At this point, since A is on the top of the stack, and since A can derive only strings that begin with *c* or *d*, an error can be detected. The message might be:

```
Unexpected b
```
or

```
Expecting c or d
```

Recovery is similar to the LR case. We need to add or delete something from the stack or input, and it is necessary to look at the various cases. One possibility is to replace *b* with *c* or *d*. The exercises explore this in more detail.

6.3.3 Recursive Descent Error Detection and Recovery

Detecting errors in recursive descent parsing is reasonably simple since the parser is in the code. For example, consider procedure factor from Chapter 4:

```
PROCEDURE Factor
  {F → (E) | Id}
  BEGIN {Factor}
  CASE nextsymbol is
"(" : PRINT ("(found")
    advance past it
    call Expression
    IF nextsymbol = ")" THEN
      PRINT (") found")
      advance past it
      PRINT ("F found")
    ELSE
      error
"Id" : PRINT ("Id found")
    advance past it
    PRINT ("F found")
  otherwise : error
  END {Factor}
```

Here, when something unexpected arrives, it is detected immediately, but no recovery is performed.

Because the parser may be many levels "down" in the recursion, recovery is difficult. The method described here is that of Wirth [1976]. It is easiest to recover, or unwind, from the recursive calls at the beginning and end of procedures. Consider a generic recursive descent procedure implementing the production N → α:

```
PROCEDURE N
  {implements N → α}
  BEGIN {N}
  . . . . . . .
  Code for α
  . . . . . . .
  END {N}
```

Certainly, the first token returned by the lexical analyzer in the code for α should be in FIRST(α). Thus, on entering procedure N, a test can be made, and if the next token is not in FIRST(α), tokens can be skipped until it is found. Similarly, by the time the procedure is exited, the next token should be FOLLOW(N). If it is not, tokens can be skipped until FOLLOW(N) or some special token such as a delimiter occurs:

```
PROCEDURE N
  {Implements N → α}
  BEGIN {N}
    Test (FIRST(α))
..
Code for α
.......
    Test (Follow (N))
  END {N}
```

where test both tests and recovers:

```
Test (ValidSet)
IF NextToken NOT IN ValidSet
  THEN Error
    WHILE NextToken NOT IN ValidSet DO
      NextToken := GetToken
    ENDWHILE
  ENDIF
```

Note that after *Code for α* has been executed, *NextToken* is possibly different from what it was before *Code for α* was executed since *Code for α* may call for tokens as it is parsing.

Improvements can still be made to this. What if the error here is that FIRST(α) has been omitted? The parse can still be continued if we allow for that fact and search also for a member of FOLLOW(N) and if found return to the procedure which called N, "pretending" to have found an N.

```
PROCEDURE N
  {Implements N → α}
  BEGIN {N}
    Test (FIRST(α) U FOLLOW(N))
    IF NextToken IN FIRST(α) THEN
      ......
      Code for α
      ......
      Test (Follow (N))
    ENDIF
  END {N}
```

We can make still more improvements here. Perhaps the error is that FOLLOWer of N (FOLLOW(N)) has been omitted also. We can continue the parse if we find a FOLLOWer of the caller of N or some special symbol that will allow the caller to continue or exit gracefully. In this case, we need to pass the caller's FOLLOW set as a parameter:

```
PROCEDURE N (FollowSet : SetOfTokens)
{Implements N → α}
BEGIN {N}
    Test(FIRST(α) U FollowSet)
    IF NextToken IN FIRST(α) THEN
       . . . . . . .
       Code for α
       . . . . . . .
       Test (FollowSet)
    ENDIF
END {N}
```

A caller of N now has the responsibility of passing an appropriate follow set for N, and if a procedure is called from within the *Code for α*, the appropriate follow set, that is, one that can follow the nonterminal represented by the called procedure, must be computed.

Example 6 shows procedure factor and how it might be called from term. Note that procedure T passes its own follow set unioned with "*" because F can be followed by "*".

```
PROCEDURE Term(FollowSet : SetOfTokens)
. . . . . . . .
Factor(FollowSet U {*})
. . .
END {Term}
```

EXAMPLE 6   Procedure factor as called from procedure term

Procedure factor (FollowSet : SetOfToken)

```
{F → (E) | Id}
BEGIN {F}
  Test({(, Id})
  IF NextToken IN {(,Id}
  THEN
  CASE nextsymbol is
"(":
  PRINT ("(found")
  advance past it
  call Expression (FollowSet U { ) } )
  IF nextsymbol = ")" THEN
    PRINT (") found")
    advance past it
    PRINT ("F found")
  ELSE
    error
  ENDIF
```

```
"Id" :
   PRINT ("Id found")
   advance past it
   PRINT ("F found")
otherwise : error
   ENDCASE
   Test (FollowSet)
   ENDIF
 END {F}
```

In the CASE statement with selector "(", the symbol ")" is added to the follow set when E is called since ")" is in the follow set for E.

6.3.4 Error Repair

Error *recovery* attempts to fix the error enough so the parse can continue in order to find any more errors. In (rare) situations, the compiler may attempt to fix the error enough so that the parse can continue *and* output code. This is termed error *repair*.

6.3.5 Programming Environments

Some programming environments (such as Turbo Pascal) come with their own editor, in addition to the Pascal compiler. Such environments are termed *integrated*. When an error is encountered, the compiler passes the user back to the editor immediately to allow her to fix the error. Thus there is no need for the compiler to recover in order to find more errors.

Syntax-directed editors also aid error handling by guiding the programmer to use the correct syntax. Such editors show the outline of the desired construct, say an IF-THEN-ELSE statement, using correct syntax.

6.3.6 Error Productions

Some errors are so common that they are almost predictable. Beginning Pascal programmers sometimes put a ";" before the ELSE in IF-THEN-ELSE statements or omit the END following a sequence of statements. It is easy to add productions to the grammar to "find" these situations.

EXAMPLE 7 Error productions

```
Program → BEGIN Statement END .
        | BEGIN Statement {Error Production}

Statement → IF Condition THEN Statement ELSE Statement
          | IF Condition THEN Statement ; ELSE Statement
          {Error Production}
```

In Example 7, two error productions are shown. This will allow the parse to continue if the programmer inadvertently makes one of these errors. During the parse, a message should be emitted if one of these erroneous situations occurs.

6.4 Summary

This chapter has discussed the four aspects of error handling: creation, detecting, reporting and recovering.

Certain language designs and constructs contribute to the creation of errors. Because some errors are so common, it is easy to predict a variety of common errors. A missing delimiter is a very common error.

Because of this ease of prediction, error productions are sometimes added to grammars. These are productions that detect common errors. Appropriate messages are emitted when erroneous input is parsed using these productions. Recovery is simple since the error is part of the language description, and the parse simply continues. The parser generator YACC has facilities for adding error productions.

The method used to detect an error varies with the type of parser, but, for table-driven parsers, an error routine is called when the top of the stack and the current input do not allow the parser to continue.

Recovery is always heuristic; that is, the action taken may not allow the parse to proceed. Further errors may be missed or spurious errors (errors that aren't really there) may be introduced. If perfect recovery were possible, then it would be possible to correct the program and emit code for it, an unlikely event.

Error reporting is usually done from a specially written error routine that is called when an error occurs. Parameters representing the place the error occurred and what the error is are passed to patterns of error messages.

Not all syntax errors are discussed in this chapter. Chapter 2 discusses errors found by the lexical analyzer. Chapter 7 will discuss further syntax errors that can be found by the semantic analyzer.

EXERCISES

1. Error Creation: For your favorite language, list common errors that you and others make.

2. Error Reporting: Make a list of schema that an error handler might include. For example,

   "At *Source*, *What* is expected"

   is a useful error schema.

3. Error Detection and Recovery: The blank entries in the table in Section 6.3.1 represent error situations. Analyze each blank entry and
   (a) invent an appropriate error message.
   (b) suggest a recovery by adding or deleting from the input or stack.

4. Error Detection and Recovery: The blank entries in the parse tables in Chapters 4 and 5 represent error situations. Analyze the entries from the tables in Section 4.3 (top-down parsing) and Section 5.2 (bottom-up parsing), and state
   (a) an appropriate error message and
   (b) possibilities to allow recovery (by adding or deleting from input, or adding or deleting from the stack).

5. LL Recovery: The method showed for recursive descent recovery can be used to determine how to recover in LL parsing.
   (a) Write an algorithm that will compute a reasonable recovery to be invoked by the error entries in the table using the current nonterminal and terminal entry. It will involve pushing or popping entries from the stack, or adding or removing symbols from the input.
   (b) Apply the algorithm in (a) to the table in Example 3 from Chapter 4.

6. Global Correction: Given an incorrect string x and a grammar G there are algorithms which find a parse for the closest string y such that the number of insertions, deletions, and changes required to change x to y is minimal. The error detector that produced the message about the reserved word misspelling in Section 6.2 uses such an algorithm. Again, such algorithms are costly in time and space, but produce nice results. Give some thought to how these algorithms might be written.

7. Error Detection in Operator Precedence Parsing: If an operator precedence parser finds no entry in the table for an input symbol and the top of the stack, then that is an error. Analyze the situations where there are blanks in the operator precedence table for the expression grammar from Chapter 5 and determine appropriate error messages as well as a recovery strategy.

8. Error Detection in Operator Precedence Parsing (cont.): If the string between <• and •> is not the right-hand side of a production, an error has occurred, requiring both a message and a recovery strategy. Input an incorrect expression string using the operator precedence table from Chapter 5 and determine some appropriate error messages and recoveries.

Compiler Project Part V   *** Optional ***
Error Handling

This assignment is suggested as optional because it is very time-consuming and may not be appropriate for a course with severe time constraints.

As this chapter suggests, the method used to handle errors is very much connected to the parsing method. Choose the section below which corresponds to your parsing method. In all cases, once a list of error messages has been developed, write an error handler that schematizes the messages as described in Section 6.2. In all cases, run your (evolving) compiler on (a) the empty program and (b) a number of other examples that will show your error handler.

Recursive Descent

Section 6.3.3 describes the general method. One can compute the FIRST and FOLLOW sets automatically, or, alternatively, since the grammar is still quite small, these sets may be determined by hand. Since this is quite a time-consuming task, some shortcuts may be in order. Not all procedures need the error routines. Procedures that merely call other procedures may be bypassed.

Implement the error recovery in a few procedures to begin with, and then, as time allows, add more.

LL Parsing or Parsing Using a Tool (Method 1)

This includes parser generators as well as projects which generate LL tables themselves. Create a list of productions that represent the syntax of possible errors. Integrate these into the grammar so that if a string is input, the parse will continue as much as possible. If your parser generator allows action routines to be added, use these to issue the appropriate messages. Some parser generators, e.g., YACC, have facilities for adding error productions.

Parsing Using a Tool (Method 2)

Some parser generators output the "table" in program form. The documentation for your tool should state this. If the table is in program form, it is possible to edit it, adding error detection and recovery as described in Sections 4.1 and 4.2.

7

Semantic Analysis

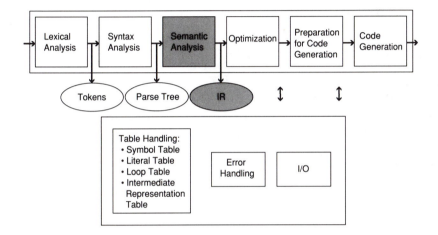

7.0 Introduction

Computer programs, like English sentences, have both a syntactic structure, illustrated by a parse tree, and a semantic structure which describes the meaning. Although there are many formal definitions of program semantics, we will take the operational semantics point of view. This approach defines the *meaning* of a program to be the executable code it generates. For us, this is the assembly language code. Thus, the actions performed by the semantic analysis phase are a beginning of the process which will generate code.

Specifically, semantic analysis performs two major actions: (1) it finishes the syntax analysis and also performs actions such as symbol table creation and (2) it translates the parse tree to an intermediate representation more appropriate for the later phases of optimization and code generation.

7.1 Static Checking

The term static checking refers to error checks made at compile time. The opposite of static is dynamic, which refers to the time a program is executing.

The error handling described in Chapter 6 finds static errors which do not correspond to the BNF description of that language. BNF is a metalanguage. A metalanguage is a way of describing another language. Thus, BNF describes a class of languages called context-free. Context-free languages can describe programming language syntactic structures such as statements and loop structures. Thus, if an END is left out, the error routines described in Chapter 6 will discover this.

However, context-free languages cannot describe the fact that a variable has been used but not defined or that a referenced label is not there. This requires contextual constraints which context-free grammars cannot specify.

There is a widely-accepted formalism, attribute grammars, for describing the range of semantic actions needed to do static checking of programming languages.

7.2 Attribute Grammars

An attribute grammar is a context-free grammar with the addition of attributes and attribute evaluation rules called semantic functions. Thus, an attribute grammar can specify both semantics and syntax while BNF specifies only the syntax.

7.2.1 Attributes

Attributes are variables to which values are assigned. Each attribute variable is associated with one or more nonterminals or terminals of the grammar.

In the grammar of Section 7.2.2, the attribute *Value* is associated with the nonterminals E, T, and F as well as with the terminal Constant. For nonterminal E, this is written:

E.*Value*

This notation indicates that nonterminal "E" has an attribute called *Value*. Attribute names will be italicized.

7.2.2 Semantic Functions

Although it is possible to evaluate some attributes at parse time, we will assume that all attributes are evaluated after the program has been parsed. Values are assigned to local attributes by equations called semantic functions. Local attributes are those which fall within the scope of a production as it appears in the parse tree. For example:

Syntax   **Semantics**
$E_0 \rightarrow E_1 + T$ $E_0.Value := E_1.Value + T.Value$

Here, the production is $E \rightarrow E + T$. The subscripts are used only to distinguish the two E's. The attribute *Value* is associated with both E's and with T. If this production were in a parse tree, it would be denoted:

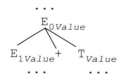

The value of the attribute *Value* is passed *up* the parse tree because $E_1.Value$ and T.*Value* are used to compute $E_0.Value$. Such attributes are termed *synthesized attributes*. Attributes whose values are passed down the tree are called *inherited attributes*.

Terminals may have only synthesized attributes, and their values are assigned by the lexical analyzer. Inherited values of the Start symbol, if any, are given values by way of parameters when attribute evaluation begins.

7.2.3   A Word about Example 1

It is common in computer science to use examples which explain the idea clearly, but which are not necessarily the best application of the idea. Thus, recursion is often introduced using factorial, a nice clear example, but not necessarily the best way to compute a factorial (iteratively is faster and takes up less space). So it is with our first example of an attribute grammar. This example computes the value of the expression represented in a parse tree. Most compilers would not do this since their ultimate goal is to put out code (which then gets executed).

EXAMPLE 1   Using attributes to evaluate expressions

Consider the following attribute grammar:

Syntax	Semantics
$E_0 \rightarrow E_1 + T$	$E_0.Value := E_1.Value + T.Value$
$E \rightarrow T$	$E.Value := T.Value$
$T_0 \rightarrow T_1 * F$	$T_0.Value := T_1.Value * F.Value$
$T \rightarrow F$	$T.Value := F.Value$
$F \rightarrow (E)$	$F.Value := E.Value$
$F \rightarrow Constant$	$F.Value := Constant.Value$

There is one attribute, *Value,* and it is a synthesized attribute. The value of this attribute can be calculated by one pass up the parse tree.

Consider the string *4 + 2 * 3* and its parse tree:

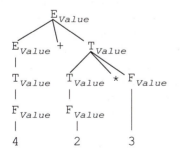

Using the semantic functions, and starting at the bottom of the tree, the various values of *Value* are computed. The lexical analyzer finds the value of Constant.*value*. This is the value shown in the above parse tree.

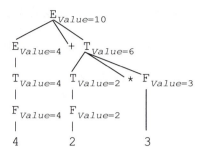

7.2.4   A More Practical Example

Example 2 is a more practical one; it might be used by a symbol table routine to attach type information to the variables in a declaration.

EXAMPLE 2   Assigning declaration types to variables

Consider the following declarations:

```
Real    A,B,Q
Integer X,Y,Q
```

and the following grammar:

```
Declaration      → Type List
List             → Variable
List             → List , Variable
Type             → Real | Integer
```

Adding attributes,

Declaration→	Type List	List.*Class* := Type.*Class*
List	→ Variable	Variable.*Class* := List.*Class*
List$_0$	→ List$_1$, Variable	List$_1$.*Class* := List$_0$.*Class*
		Variable.*Class* := List$_0$.*Class*
Type	→ Real | Integer	Type.*Class* := *LexValue*(Type)

where *LexValue* is the value of the token (Real or Integer) found by the lexical analyzer. This attribute, *Class,* for the nonterminal Type in the last production, is said to be *intrinsic* because its value is given, not computed. There is again only one attribute, *Class*. It is an inherited attribute since its value is computed while going down the tree. This can be seen by noticing that the semantic functions compute the attribute for the grammar symbols on the right-hand side of production in terms of the values on the left-hand side. For example, in the semantic function:

```
Variable.Class := List.Class
```

the nonterminal "Variable" appears on the right-hand side of the production:

```
List → Variable
```

while the nonterminal "List" appears on the left-hand side.

The parse tree before attribute evaluation for the first line above is:

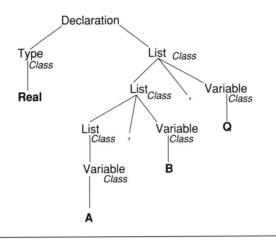

7.2.5 Other Static Checks

Other static checks include type checking expressions to make sure operators are compatible with their operands, checking that a variable used has been declared, checking that an array declared to be of a certain dimension is used with the correct number of subscripts, and a myriad of other tasks. Some static checks are language dependent since not all languages have the same constraints. FORTRAN, for instance, does not require variables to be declared before being used.

7.3 Translation to an Intermediate Representation

In order to perform optimizations on a program and to generate code, it is convenient to produce an intermediate representation better suited to these tasks than the parse tree representation. Chapter 1 shows an intermediate representation, abstract syntax trees (AST's), used in many modern compilers. Diana, the intermediate language for Ada, is an attributed AST.

We will show the following intermediate representations: postfix notation, abstract syntax trees (again), and three-address code.

7.3.1 Polish Postfix

Polish postfix is a linear line of code which is more useful for code generation than for the optimization phases since it is difficult to do the sorts of transformations on it which are performed during the optimization phase.

Anyone who has ever used a Hewlett Packard calculator is familiar with Polish postfix (for expressions). Essentially, Polish postfix puts the operands first followed by their operator.

Thus,

    A + B

becomes

$A\ B\ +$

and

$A\ +\ B\ *\ C$

becomes

$A\ B\ C\ *\ +$

while

$A\ *\ B\ +\ C$

becomes

$A\ B\ *\ C\ +$

The assignment statement:

$S\ :=\ A\ +\ B\ *\ C$

is denoted

$S\ A\ B\ C\ *\ +\ :=$

It is more difficult to devise a reasonable postfix notation for other programming language constructs such as IF statements.

7.3.2   Abstract Syntax Tree (AST)

An abstract syntax tree is a parse tree stripped of unnecessary information such as singleton productions (e.g., E→T→F). Each nonleaf represents an operator and each leaf represents an operand.

Thus $A + B$ might have the following parse tree and abstract syntax tree:

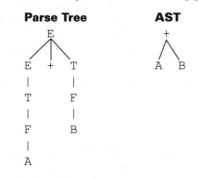

$S := A + B * C$ has abstract syntax tree:

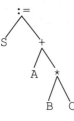

which can be represented in parenthesized form as

$$(:= ("S" + ("A" * ("B" "C"))))$$

The parenthesized form is often used by the compiler designer for debugging output. It is also the recommended output for the part of the project described at the end of this chapter. It is essentially a preorder transversal of the tree.

IF A < B THEN Max := B is represented

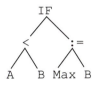

and *A[I] := B* is

Even though AST's are more convenient for optimization, there is a one-one correspondence between AST's and Polish postfix:

Thus, AST's can be easily converted to Polish prefix if desired. The "tree" can be seen by starting at the right-hand end and taking the upper of the two arrows as the left child and the lower as the right child.

7.3.3 Three-Address Code

This intermediate form is called *three-address* because each "line" of code contains one operator and up to three operands, represented as addresses. Since most assembly languages represent a single operation in an instruction, three-address code is closer to the target code than the parse tree representation. There are a number of variants of three-address code, some more appropriate than others for optimization.

Quadruples (a Three-Address Code)

Quadruples ("quads" for short) consist of an operator (up to), two operands and a result.

A + B would translated into quads as:

$A + B * C$ would be translated into two quads:

```
*   B   C   T1
+   A   T1  T2
```

and $S := A + B * C$ would be:

```
*       B   C   T1
+       A   T1  T2
:=      T2  __  S
```

Notice that the last quad here had only one operand rather than two.

An alternative notation for quads, one we will use in the chapters on optimization, is to write them as a sequence of assignment statements. Thus $S := A + B * C$ would be written:

```
T1 := B * C
T2 := A + T1
S  := T2
```

If we think of *GOTO L* as an operator and a result, the quad would be:

```
GOTO  __  __  L
  ↑            ↑
Operation   Result
```

$A[I] := B$ would be:

```
[ ]   A   I   T1
:=    T1      B
```

The reasoning for the intermediate code for IF statements is similar to the reasoning used to implement such constructs into assembly language. For example, *IF A < B THEN Max := B* would be:

```
<       A   B   Label 1
GOTO            Label 2
                Label 1
:=      B       Max
                Label 2
```

Here, the first line can be thought of as "If A < B then execute the code at Label 1". The second line can be thought of as "Otherwise execute the code at Label 2". The third line has only a result field—Label 1. These lines can be translated easily into assembly language code.

Triples (A Three-Address Code)

Quadruples use a name, sometimes called a temporary name or "temp", to represent the single operation. Triples are a form of three-address code which do not use an extra temporary variable; when a reference to another triple's value is needed, a pointer to that triple is used. We show the same examples as above for triples:

$A + B$ would be translated into triples as:

```
    +           A  B
    ↑           ↑  ↑
Operation    Operands
```

$A + B * C$ would be translated into two triples:

```
(1)   *   B   C
(2)   +   A   (1)
```

Here, the triples have been numbered and a reference to the first triple is used as an operand of the second. $S := A + B * C$ would be

```
(1)   *    B    C
(2)   +    A    (1)
(3)   :=   S    (2)
```

Triples are really an abstract syntax tree in disguise:

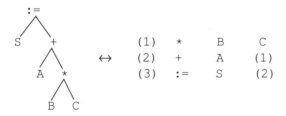

Triples are difficult to optimize because optimization involves moving intermediate code. When a triple is moved, any other triple referring to it must be updated also. A variant of triples called indirect triples is easier to optimize.

Indirect Triples (A Three-Address Code)

Here, the triples are separated from their execution order:

Triples:
```
1.    *    B    C
2.    +    A    (1)
3.    :=   S    (2)
```

Execution order: 1, 2, 3

Since the execution order here is the same as the order of the triples themselves, it is difficult to see the use of indirect triples. Example 3 shows a case where the execution order is not the same as the order of the indirect triple list.

With indirect triples, optimization can change the execution order, rather than the triples themselves, so few references need be changed.

EXAMPLE 3   Optimizing indirect triples

```
for i := 1 to 10 do        ↔        a := b * c
   begin              (optimize)     for i:= 1 to 10 do
      a := b * c;                       begin
      d := i * 3;                          d := i * 3
   end                                  end
      •                                    •
      •                                    •
   end;                                end;
```

The triples for the original unoptimized version might be:

```
(1)  :=   1   i
(2)  *    b   c
(3)  :=  (2)  a
(4)  *    3   i    Execution Order : 1 2 3 4 5 6 7 8
(5)  :=  (4)  d
(6)  +    1   i
(7)  LE   I   10
(8)  IFT  GO (2)
```

If the computation $a := b * c$ is moved out of the loop, as in the right-hand optimized version, the execution order is 2 3 1 4 5 6 7 8. Notice that only the reference in (8) need be changed.

As an indication of how intermediate form might be generated during parsing, Section 7.3.6 describes how to generate abstract syntax trees during recursive descent parsing. Similar (semantic) routines would be written for generating AST's if the parser were generated using a tool. Most tools allow such routines to be written.

7.3.4   Other Intermediate Representations

Strictly speaking, any form between the source program and the object program might be called an intermediate representation. The optimization process often creates a graphical representation called a control flow graph (see Lemone, 1992, Chapters 7 to 9). We have discussed attributes only briefly, but a parse tree or an abstract syntax tree with attached attributes is often called an attribute tree or a semantic tree—yet another possible intermediate form.

7.3.5   Arrays

Array references are often translated to intermediate code with an array operator and two children. The left child is the name of the array, and the right child is the subscript expression.

Thus,

```
Temp := List[I]
```

as an abstract syntax tree would be

and as quadruples (using the assignment statement notation) would be:

```
T1 := List[I]
Temp := T1
```

Using this notation, the code generator would have to compute the array offset for machines which do not have an indexing addressing mode. The alternative is to expand the intermediate representation.

Using "( )" to mean "contents of" and "addr" to mean "address of," the quadruples for "Temp := List[i]" would be:

```
T1 := addr (List)
T2 := T1 + I
Temp := (T2)
```

Similarly, if the array reference is on the left-hand side of an assignment statement as in "List[I] := Temp", low-level quadruples would be:

```
T1 := addr (List)
T2 := T1 + I
(T2) := Temp
```

Even if a machine has an indexing addressing mode, translating array references into their low-level format may allow optimizations (for example, if the subscript reference were a common subexpression).

7.3.6 Adding AST Routines to Recursive Descent Parsing

An abstract syntax tree node consists of an information field and a number of pointer fields. We will show an example for the expression grammar with assignment statements added and presume a binary tree so that there are two pointer fields to a left and right child in addition to the information field.

Thus if an AST node is called *Node*, these three fields will be denoted as *Info(Node)*, *Left(Node)* and *Right(Node)*. *Get(Node)* will create a new empty AST node.

The method consists of adding a parameter to each of the procedures which will carry the (partial) AST from procedure to procedure, adding on to it as appropriate.

Consider the recursive descent procedure for an assignment statement:

```
{Assignment → Variable := Expression}

PROCEDURE Assignment (Tree : AST)
BEGIN {Assignment}
   .
   .
   .
```

```
IF NextToken = Variable THEN
  Get (Node)
  Info(Node) := Variable
  Left(Node) := NIL
  Right(Node) := NIL
  Tree := Node
IF NextToken = ":=" THEN
  Get (Node)
  Info(Node) := ":="
  Left(Node) := Tree
  Right(Node) := NIL
  Tree := Node
  Expression (Right(Tree)) ; Returns a (pointer to a)
                           ; right subtree
        •
        •
        •

END {Assignment}
```

Following this pseudo-code for the assignment statement

```
A := B
```

we obtain the following abstract syntax tree, presuming that the call to *Expression (Right(Tree))* returns a node whose information field contains *B* and *Right(Tree)* is a pointer to it.

To show this, we will continue this process for Expression, Term and Factor (in outline form):

```
{Expression → Term {+Term}}

PROCEDURE  Expression (Tree: AST)
BEGIN {Expression}
       •
       •
    Term(Tree)
    WHILE NextToken = "+" DO
      Get (Node)
      Info(Node) := "+"
      Left(Node) := Tree
      Term(Right(Node))
      Tree := Node
       •
       •
       •

END {Expression}
```

```
{Term → Factor {* Factor}}

PROCEDURE Term (Tree : AST)
BEGIN {Term}
     .
     .
     .
   Factor (Tree)
   WHILE NextToken = "*" DO
     Get(Node)
     Info(Node) := "*"
     Left(Node) := Tree
     Factor(Right(Node))
     Tree := Node
     .
     .
     .

END {Term}

{Factor → Const | (Expression)}

PROCEDURE Factor (Tree : AST)
BEGIN {Factor}
     .
     .
     .
  IF NextToken <> "(" THEN
     Get(Node)
     Info(Node) := Token
     Right(Node) := NIL
     Left(Node) := NIL
     Tree := Node
     .
     .
     .

END {Factor}
```

The reader is invited to trace this for the expression $A := B$ (see Exercise 9).

7.4 Summary

This chapter is a description of the semantic analysis phase of compiling. It is an intermediate phase linking the front-end analysis phase of a compiler with the back-end synthesis phase. The specification language, an attribute grammar, is introduced. Other specifications may also be used. There is also a lot more to learn about attribute grammars and the evaluation of attributes. Lemone (1992), Chapter 6, covers this subject.

The next chapter, Symbol Tables, also describes activities which come under the umbrella of semantic analysis.

EXERCISES

1. Translate the statement $A := 2 * (Bee + Cee/Dee)$ into:
   (a) a parse tree using a reasonable grammar for assignment statements
   (b) an abstract syntax tree
   (c) quadruples
   (d) triples
   (e) indirect triples

2. Translate the statement $A[2*i] := A[i] + 2 * i$ into:
   (a) an abstract syntax tree
   (b) quadruples
   (c) triples
   (d) indirect triples

3. Translate the statement $WHILE\ (1 < P)\ AND\ (P < 3)\ DO\ P := P + Q$ into:
   (a) an abstract syntax tree
   (b) quadruples
   (c) triples
   (d) indirect triples

4. Show an abstract syntax tree for the following expression:

   ```
   C[i * n + j] := (a + X) * y + 2 * 768 + ((L - m) * (-k)) / z
   ```

5. Consider the following (simplified) grammar for array references:

   ```
   List  → Id
   List  → Id [EList]
   EList → E
   EList → EList , E
   E     → Id
   ```

   Define an attribute called *NumberOfDimensions* and semantic functions
   which will leave the number of dimensions in an array reference at the top of
   the parse tree for List:
   (a) What might such information be used for?
   (b) Show the semantic functions which will calculate the value (that is, leave
       the value at the root):

Grammar	**Semantic Functions**
`List  → Id`	
`List  → Id [EList]`	
`EList → E`	
`EList → EList , E`	
`E     → Id`	

   (c) Is *NumberOfDimensions* an inherited or synthesized attribute?
   (d) Show a parse tree and attribute evaluation for $A[I, J, K]$.

6. Modified Knuth example (Knuth, 1971a) Consider the following attribute
   grammar:

Productions   **Semantics**

0 : Number → Sign List   (i) List.*Scale* := **0**

                                        (ii) Number.*Value* :=

                                             IF Sign.*Neg* THEN - List.*Value*

                                             ELSE List.*Value*

1. Sign → **+**   (i) Sign.*Neg* := **False**

2. Sign → **-**   (i) Sign.*Neg* := **True**

3. List → BinaryDigit   (i) BinaryDigit.*Scale* := List.*Scale*

                                        (ii) List.*Value* := BinaryDigit.*Value*

4. List$_0$ → List$_1$ BinaryDigit   (i) List$_1$.*Scale* := List$_0$.*Scale* + 1

                                        (ii) BinaryDigit.*Scale* := List$_0$.*Scale*

                                        (iii) List$_0$.*Value* := List$_1$.*Value*

                                                            + BinaryDigit.*Value*

5. BinaryDigit → **0**   (i) BinaryDigit.Value = **0**

6. BinaryDigit → **1**   (i) BinaryDigit.Value = **2**$^{BinaryDigit.Scale}$

(a) Identify the attributes and classify them as inherited, synthesized or intrinsic.

(b) Parse and evaluate the attributes for the string: *–1 0 1.*

(c) What do these attributes do?

7. Consider the following attributed grammar for a program consisting of assignment statements. The minus sign, –, in (iv) and (v) of production 3 represents set difference.

Productions   **Semantics**

1. Program → BEGIN Statement END   (i) Statement.*Live* := Statement.*Use*

2. Statement → Assignment   (i) Statement.*Use* := Assignment.*Use*

                                        (ii) Statement.*Defn* :=

                                                      Assignment.*Defn*

                                        (iii) Assignment.*Live* :=

                                                      Statement.*Live*

3. Assignment → *Variable* := Expression (i) Expression.*Use* := {Variables in

                                                              Expression}

                                        (ii) *Variable.Defn* :=

                                                      {LexValue(*Variable*)}

                                        (iii) Assignment.*Defn* :=

                                                      Variable.Defn

                                        (iv) Assignment.*Use* := Expression.*Use*

                                                      - *Variable.Defn*

                                        (v) *Variable.Live* := Assignment.*Live*

                                                      - *Variable.Live*

                                        (vi) Expression.*Live* :=

                                                      Expression.*Use*

where Expression can be any valid expression, e.g., *A + B*.

(a) List the synthesized and inherited attributes.

(b) Parse and evaluate attributes for the program:

```
BEGIN
   B := A + B
END
```

8. Translate the two-line example from Chapter 1

```
X1 := a + bb * 12;
X2 := a/2 + bb * 12;
```

   into

   (a) Postfix
   (b) Quads
   (c) Triples

9. Use the procedures in Section 7.3.6 to create abstract syntax trees for

   (a) A + B
   (b) A + B * C

10. In this chapter, we did not distinguish between addresses and contents. For example, in the assignment statement

```
A := A,
```

   the *A* on the left is an address while the *A* on the right represents a value. If we denote assignments by

```
l-value := r-value
```

   and

   addr (A) to denote A's l-value,

   and

   * A to denote A's r-value,

   then quads for *A[I]* := *B* would be written, using the alternative notation:

```
T1 := addr (A)
T2 := I
T3 := T1 + T2
T4 := B
*T3 := T4
```

   Show similar quads for the expressions of Exercise 8.

11. Translation Grammars: A translation grammar is an attribute grammar with two kinds of terminal symbols, *input symbols* and *action symbols*. The following grammar can be used to translate the infix expression $c * (a + b)$ to the postfix expression *{c} {a} {b} {+} {*}*.

```
E → E + T   {+}  ←──────────┐
E → T                       │
T → T * F   {*}             │
T → F              Output Symbol
F → (E)
F → a       {a}
F → b       {b}
F → c       {c}
```

For our example . . .

```
c * (a + b)
```

the parse tree is:

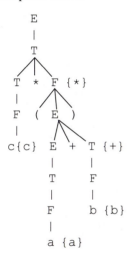

The input string can be read by reading leaves not within set braces, "{}", while the translated string can be read by reading the symbols *within* the braces, "{}".

Write a translation grammar which will generate octal strings and translate them at the same time to their equivalent binary form.

12. (a) Consider the following attribute grammar:

BNF	Semantic Functions
$N_1 \rightarrow N_2$ c	$N_1 . CNum = N_2 . CNum + 1$
	$N_1 . DNum = N_2 . DNum$
$N_1 \rightarrow N_2$ d	$N_1 . CNum = N_2 . CNum$
	$N_1 . DNum = N_2 . DNum + 1$
$N \rightarrow$ d	$N . CNum = 0$
	$N . DNum = 1$
$N \rightarrow$ c	$N . CNum = 1$
	$N . DNum = 0$

(b) Parse and compute attribute values for the string *c d c*.

Compiler Project Part VI

Abstract Syntax Trees

Define a set of abstract syntax tree nodes for the language which your compiler currently implements (see BNF in Chapters 3 and 5).

EXAMPLE 1

```
        plus
       /\
  opnd1  opnd2
```

is an AST node for *a + b* where *a* is opnd1 and *b* is opnd2. In parenthesized form, this could be written:

```
(plus ("a" "b")) or (+ ("a" "b"))
```

EXAMPLE 2

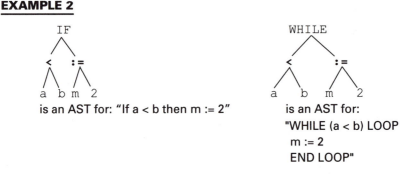

is an AST for: "If a < b then m := 2"

is an AST for:
"WHILE (a < b) LOOP
m := 2
END LOOP"

In parenthesized form these might be written:

```
(IF ("<" ("a" "b") ":=" ("m" "2"))) and
(WHILE ("<" ("a" "b") ":=" ("m" "2")))
```

or, with some indenting,

```
(IF
   ("<" ("a" "b")
   ":=" ("m" "2"))) and
(WHILE
   ("<" ("a" "b")
   ":=" ("m" "2")))
```

(Other indentings are possible here.)

Two of the productions that relate to a node such as shown in Example 1 are:

```
...
Relation          → SimpleExpression
SimpleExpression → Term₁ {AddingOperator Term₂}
...
```

(The dots represent the productions which come before and after these two.)

If we create a data structure:

then we can add attributes such as the following:

```
...
Relation              → SimpleExpression
                        Relation NodePtr = SimpleExpression.NodePtr

SimpleExpression  → Term₁ {AddingOperator Term₂}
                        SimpleExpression.NodePtr := {GetNode}
                        SimpleExpression.Info := "+"
                        SimpleExpression.Left := Term₁.NodePtr
                        SimpleExpression.Right := Term₂.NodePtr
```

Attributes and semantic functions may be added analogously to those of the "+" node above.

Design attributes and semantic functions which will translate your programs to abstract syntax trees. Add semantic actions to whatever tool you are using to translate input programs to AST's. Print them out in parenthesized form. The following is sample output; line numbers are optional.

```
Unit [
 Line (1)
 (BeginStmt)
 Line (2)
 (":="
   ("a", "b3"))
 (":="
  Line (3-4)
  ("xyz", "+"
      ("a" "+"
        ("b" "-"
          ("c" "/"
            ("p" "q")))))
 Line (5)
 (":="
   ("a" "*"
     ("xyz" "+"
       ("p" "q"))))
 Line (6)
 (":="
  ("p" "-"
    ("a" "-"
      ("xyz" "p"))))
 Line (7)
 (EndStmt) ]
```

Method 1 If your compiler uses recursive descent parsing, use the methods described in Section 7.3.6, printing out the AST in parenthesized form.

    or

Method 2 Add semantic actions to whatever tool you are using to translate input programs to parenthesized AST's. This will require **you** to write the routines which actually output the AST's. Reading the routines added to the recursive descent parser in Section 7.3.6 will help.

Run your program with (a) the program of Project, Part I, (b) the programs of Project, Part III and IV and (c) a program of your choice. (You need no longer print out tokens and parse information.)

8

Symbol Tables

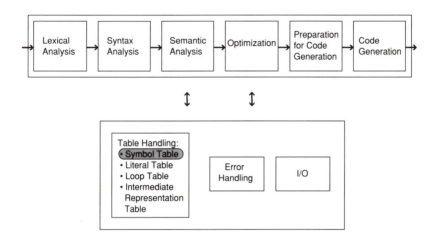

8.0 Introduction

The symbol table records information about each *symbol name* in a program. Historically, names were called symbols and hence we talk about a symbol table rather than a name table. In this chapter, the word symbol will mean *name*. The semantic analysis phase creates the symbol table since it is not until semantic analysis that enough information is known about a name to describe it. Code generation uses the symbol table to output assembler directives of the appropriate size and type.

It is important to distinguish between a *symbol* and an *identifier* since the same identifier may represent more than one name. For example, in FORTRAN, one can write:

```
COMMON /X/  X
F(X) = X + 1
```

Here, the single identifier *X* represents three names: (1) the name of a common block (a shared section of memory) in */X/*, (2) the element *X* to be found there and (3) the dummy variable in a function definition.

Another example is found in block structured languages:

The picture implies that there is an outer block (or procedure) with a declaration of *x* and two inner blocks (or procedures) each with its own declaration of *x*; thus, the single identifier *x* again represents three different names or symbols.

Many compilers set up a table at lexical analysis time and fill in information about the symbol later during semantic analysis when more information about the variable is known. A classic example comes from the syntax used to refer to arrays and functions in FORTRAN and Ada. In both of these languages, F(2) might refer to an element F_2 of an array F or the value of function F computed using argument 2. For the lexical analyzer to make the distinction, some syntactic and semantic analysis would need to be added.

Symbol tables provide the following information:

- Given an identifier, which *name* is it?
- What information is to be associated with a name?
- How do we associate this information with a name?
- How do we access this information?

Some symbol tables also include the keywords in the same table. The alternative is to have a separate table for keywords.

8.1   Symbol Attributes

Each piece of information associated with a name is called an attribute (not to be confused with the term *semantic attribute* from Chapter 7).

Attributes are language dependent, but might include the characters in the name, its type and even storage allocation information such as how many bytes the value will occupy. Often, the line number where the name is declared is recorded as well as the lines where the symbol is referenced. If the language contains scopes, as most do (FORTRAN is an exception), then the scope is often entered into the symbol table. We will discuss a number of these attributes separately beginning with the class attribute. The *class* of a name is an important attribute.

8.1.1   The Class and Related Attributes

A name in a program can represent a variable, a type, a constant, a parameter, a record, a record field, a procedure or function, an array, a label, or a file, to name just a few possibilities. These are values for an attribute called the symbol's *class*. Of course, not all languages have all of these possibilities—FORTRAN has no records—or they may be described using other terms—FORTRAN uses the word subroutine instead of procedure.

Once a name's class is known, the rest of the information may vary depending upon the value of class. For example, for a name whose class is variable, type, constant, parameter, or record field (to name a few), there is another attribute which records the names's *type*. Notice that this is somewhat recursive since *type* is also one of the possible classes. The recursion terminates with a name of some basic type. For example, in Pascal, the basic types are integer, real, character or Boolean.

For a name whose class is procedure or function, there are other attributes which indicate the *number of parameters*, the *parameters* themselves, and the *result type* for functions.

For a name whose class is array, other reasonable attributes are the *number of dimensions* and the *array bounds* (if known).

For a name whose class is file, other attributes might be the *record size*, the *record type* (sequential), etc.

Again, the possible classes for a name vary with the language, as do the other attributes (see Exercise 1).

8.1.2   Scope Attribute

Block structured languages allow declarations to be nested; that is, a name can be redefined to be of a different class. A similar problem occurs when nested procedures or packages redefine a name. The name's *scope* is limited to the block or procedure in which it is defined. In Ada, FOR-LOOP variables cause a new scope to be opened (containing only this variable).

Example 1 shows a program *Main* with global variables a and b, a procedure P with parameter x and a local variable a. Information about each of these names is kept in the symbol table. References to a, b, and x are made in procedure P while main program *Main* makes references to procedure P and to its variable a.

EXAMPLE 1   Block structure

```
PROGRAM Main
  GLOBAL   a,b
  PROCEDURE  P (PARAMETER x)
    LOCAL  a
  BEGIN {P}
    ...a...
    ...b...
    ...x...
  END {P}
  BEGIN {Main}
    Call P(a)
  END {Main}
```

The scope, perhaps represented by a number, is then an attribute for the name. An alternative technique is to have a separate symbol table for each scope.

8.1.3   Other Attributes

Section 8.1.1 described how a name's other attributes may vary according to the value of its class attribute. Section 8.1.2 emphasized the importance of a *scope* attribute.

Other attributes for names include the actual *characters* in the name's identifier, the *line number* in the source program where this name is declared and the *line numbers* where references occur.

8.1.4   Special Attributes

Special purpose languages often have special names. Object-oriented languages, for example, may have method names, class and object names, as well as the usual types. Scoping is particularly important in object-oriented languages because names often *inherit* operations from super classes which contain them.

Functional programming languages such as LISP have scoping issues which involve "binding" a name to a particular environment. These issues are outside the scope (no pun intended) of this text.

8.2 Symbol Table Operations

There are two main operations on symbol tables: (1) *Insert* (or *Enter*) and (2) *Lookup* (or *Retrieval*).

Most languages require declaration of names. When the declaration is processed, the name is inserted into the symbol table. For languages not requiring declarations, e.g., FORTRAN, a name is inserted on its first occurrence in the program. Each subsequent use of a name causes a symbol table lookup operation.

Given the characters representing a name, searching finds the attributes associated with that name. The search time depends on the data structure used to represent the symbol table.

8.3 External Data Structures for Symbol Tables

A symbol table makes a wonderful data structure example since the pros and cons for various data structures can be easily explored.

Notice, first of all, that a name is entered once, but may be retrieved many, many times. In fact, even the *enter* operation may be preceded by a *lookup* operation to ascertain that the name is not already there. Thus, data structures which search rapidly are to be preferred for efficiency.

Some data structures make it easier to implement block structure information.

We will consider various options one by one.

8.3.1 An Unordered List

An unordered list would enter each name sequentially as it is declared. The lookup operation must then search linearly and thus, in the worst case, would have to look at all n entries and in the average case at half of them. Thus the search time is of the order n, $O(n)$.

There is really no good reason to have such an inefficient data structure unless it is known that the number of entries will be exceedingly small, perhaps less than a couple of dozen names. Example 2 shows an unordered list structure for the program of Example 1.

EXAMPLE 2 An unordered list structure

Characters	Class	Scope	Declared	Referenced	Other
Main	Program	0	Line 1		
a	Variable	0	Line 2	Line 11	
b	Variable	0	Line 2	Line 7	
P	Procedure	0	Line 3	Line 11	1 parameter, x
x	Parameter	1	Line 3		
a	Variable	1	Line 4	Line 6	

Of course, there could also be a separate table for each scope.

8.3.2   An Ordered List

By ordered, we mean ordered according to the characters in the variable's name.

With an ordered array, a binary search could be done in $O(\log_2 n)$ time on the average, but the *enter* operation will take more time since elements may have to be moved. To do a binary search, however, we need to implement the ordered list as an array.

Representing the list as an ordered linked list brings us back to $O(n)$ for the *lookup* operation although the *enter* operation would, on the average, be simpler since only pointers need to be changed rather than the position of each element.

Example 3 shows an ordered list. Notice that we cannot see whether this has been implemented as an array or as a linked list.

EXAMPLE 3   An ordered list structure

Characters	Class	Scope	Other Attributes		
			Declared	Referenced	Other
a	Variable	0	Line 2	Line 11	
a	Variable	1	Line 4	Line 6	
b	Variable	0	Line 2	Line 7	
Main	Program	0	Line 1		
P	Procedure	1	Line 3	Line 11	1 parameter, x
x	Parameter	1	Line 3		

8.3.3   Binary Tree

Binary trees combine the fast search time of an ordered array, $O(\log_2 n)$ on the average, with the insertion ease of a linked list. Worst case (see Exercise 2) is still $O(n)$ for retrievals.

Example 4 shows our example for a binary tree.

EXAMPLE 4   A binary tree structure

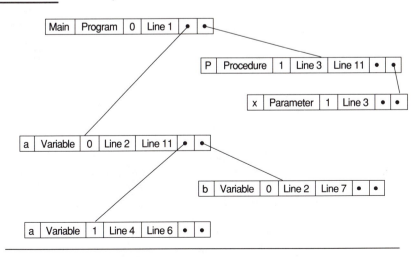

In Example 4, the tree is reasonably well balanced.

Binary trees are space-efficient since they consume an amount of space proportional to the number of nodes. Since new nodes are added as leaves, scoping is difficult unless separate trees are maintained for each scope.

Printing out an alphabetized list of names is also quite easy.

If the tree is allowed to become unbalanced, searching can degrade to a linear search. Exercise 2 discusses a worst case for a binary tree implementation. Exercise 6 discusses the use of AVL trees for symbol tables.

8.3.4 Hash Tables

For efficiency, hash tables are the best method. Most production-quality compilers use hashing for their symbol table structure.

Lots of hashing functions have been developed for names (see Exercise 5). It has been joked that the hashing function for symbol tables is irrelevant as long as it keeps the variables I, N, and A distinct!

The hashing functions consist of finding a numerical value for the identifier, perhaps some combination of the ASCII code as a number or even its bit code, and then performing some of the techniques used for hashing numbers (taking the middle values, etc.).

Example 5 creates a hash table using the formula

$$H(Id) = (\# \text{ of first letter} + \# \text{ of last letter}) \bmod 11$$

where # is the ASCII value of a letter.

EXAMPLE 5   A hash table structure

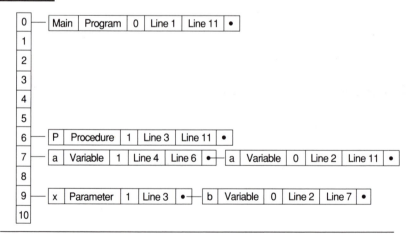

Notice in Example 5 that both instances of variable a have hashed to the same position in the hash table and that both x and b have hashed to the same position.

If the hash function distributes the names uniformly, hashing is efficient—O(1) in the average case. The worst case, where all names hash to the same number, has a time of O(n). This case is unlikely, however.

More space is generally needed in a hash symbol table structure, so it is not as space-efficient as binary trees or list structures.

If new entries are added to the front of the linked structure, scoping is easily implemented in a single table. In fact, separate tables would usually take up too much extra space

8.3.5 Other External Data Structures

Another data structure for a symbol table is a stack where a pointer is kept to the top of the stack for each block. In this structure, names are pushed onto the stack as they are encountered. When a block is completed, that portion of the stack and a pointer to it are moved so that the containing block's names can be completed. This is an easy structure to implement but relatively inefficient in operation.

Combinations of the data structures described here are common. Thus, we could have a hash structured symbol table implemented as a stack.

8.4 Internal Structure for Symbol Tables

Once the basic data structure is decided upon, we need to decide how the names and their attributes are to be stored.

The logical view of a symbol table is a list of names, each with its own list of attributes. Each entry is thus of a variable size since the number of attributes of a name depends on its class. A data structure which takes these variable length entries into account is most space efficient, and usually a symbol "table" is not a single table, but more like a data base or a collection of data structures that work together.

Thus the symbol table structure might consist of a *string* table where the actual characters in a name are stored, a *class* table where the class of a name is stored, and a *name* table consisting of pointers to the other two tables for each (different) name in the program.

Example 6 shows a possible symbol table structure for a simple declaration. The tables shown there appear to be arrays, but they could be linked lists or some other appropriate data structure.

EXAMPLE 6   Tables for the declaration   *A : Array [1..100] of Integer;*

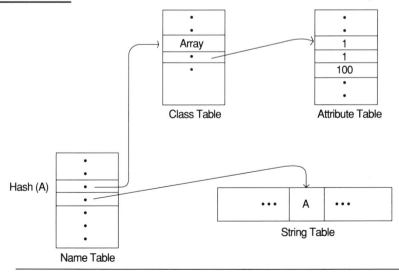

Here, the name table has two pointers—one to A's class (*Array*). A pointer from the class table points to the attribute table. The attributes there indicate that the array is one-dimensional, with a lower limit of 1 and an upper limit of 100.

The second pointer from the name table is to the actual character string, *A*. String tables are more efficient for languages such as older versions of FORTRAN where the length of an identifier is restricted to six characters.

Other attributes might include the block number or embedded procedure level where *A* is declared.

8.5   Other Symbol Table Techniques

Symbol table information or pointers to symbol table information can be attached to the nodes of the parse tree or abstract syntax tree. Nonunique names can be replaced with dummy names. A technique for dealing with embedded declarations is to have a separate symbol table for each level and to stack the symbol tables.

In a one-pass compiler, code is probably put out as soon as a block is closed. Thus symbol table structures such as a stack, which can discard the table information for nested procedures as soon as they are processed, are useful.

Attribute information is more important in a multi-pass compiler.

Sophisticated languages need sophisticated compilers and hence sophisticated symbol table structures and operations. The exercises discuss a few of these more complicated structures (e.g., records) which a symbol table must handle. Other language constructs requiring artful techniques include the Pascal WITH statement, object-oriented programming concepts such as inheritance (see Lemone, 1992, Chapter 2), the implicit declarations of languages like FORTRAN and BASIC, the IMPORT and EXPORT statements of Modula2 and polymorphism (see Lemone, 1992, Chapter 2) of Ada.

Compilers, and hence the symbol table, are usually written in a high-level language. Thus, the symbol table implementation is dependent on the language constructs found in this language. To be able to create a symbol table for large programs, but yet not waste space when creating symbol tables for small programs, requires some sort of efficient dynamic storage allocation. It has been suggested that dynamic arrays are an appropriate data structure since they avoid the "spaghetti-like" code of pointer variables and the problem of garbage collection, yet can allocate different sized objects in an efficient dynamic way.

8.6   Summary

Symbol table access can consume a major portion of compilation time. Every occurrence of an identifier in a source program requires some symbol table interaction. For a linear search this might consume as much as one-quarter of the translation time.

Symbol table actions are characterized by the fact that there are more retrievals than insertions. The entries are variable-sized depending, in particular, on the symbol's class. Thus, for efficiency of time, data structures which allow fast lookup algorithms are appropriate, and for efficiency of space data structures which allow variable-sized information are appropriate. As usual, these two issues, time vs. space, may conflict.

Most compilers use either hashing or binary trees.

Since a compiler is a large software program, often maintained by people who did not write it, good programming mandates that the symbol table be written using established software engineering techniques. The project at the end of this chapter suggests implementing the symbol table as an abstract data type.

Although not discussed in this chapter, deletion from the symbol table is another operation which should be efficient. In a one-pass compiler, the variables in a block may be discarded upon exit from the block. In a multi-pass compiler the "last" phase should be able to accomplish this same task.

There are often other tables in a compiler. For example, literal constants may be kept in a table. Keywords may also be kept in a table. If they are kept in the same table as user-defined names, they should be marked as "keywords". If a hash table is used, a hashing function which maps keywords to a different part of the table is useful.

EXERCISES

1. For your favorite high-level language, devise a set of classes and a set of other attributes appropriate for symbol table entries.

2. Section 8.3.2 indicates that the worst time behavior for a binary structured symbol table is O(n). Show a set of declarations in your favorite high-level language whose symbol table would have this worst case behavior.

3. Consider the following declaration:

```
A = record
   A1 :  array[1..10,0..9] of real;
   A2 :  boolean;
end
```

Show a possible symbol table organization.

4. The following represents a *possible* (internal) symbol table structure for a variable called *Test* in a language which has array, record, integer, real, character, and pointer data types. Using a pseudocode or a real language, show what the language declaration might look like for this symbol table entry.

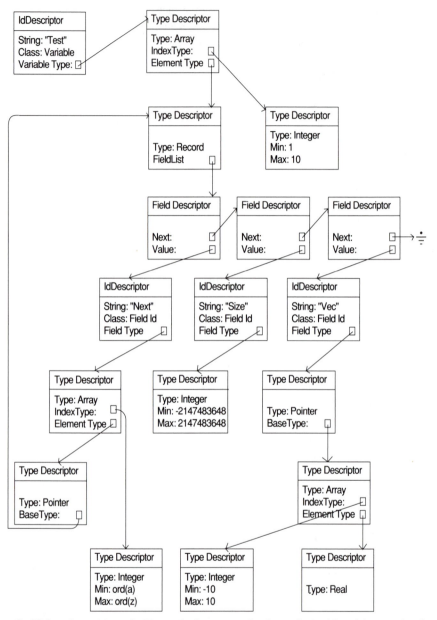

5. Using the program in Example 1, create a hash symbol table with n entries for each of the following hashing functions:

    (a) h = (# first letter + # last letter) mod n, where # is the ASCII value, for n = 10.

    (b) h = (# first letter + # last letter) mod n, for n = 13.

    (c) h = (# first letter ∗ # last letter) mod 11, where # is the ASCII value.

    (d) Repeat (a)–(c) using the first and last two letters rather than one.

    (e) Repeat (a)–(c) using all the letters in an identifier.

6. Binary trees may become unbalanced as new nodes are added (see Exercise 2). However, keeping a binary tree balanced as new nodes are added may be too time consuming. A compromise is the use of AVL trees which allow either the left or the right subtree's length to be one longer than the other. Searching and inserting (as well as deleting) are $\log_2 n$ operations. Many data structure books contain information about AVL trees. Create an AVL tree for the declaration found in Example 1. Show an addition being made to this tree.

Compiler Project Part VII
Symbol Tables, Declarations, and Arrays

This is a long assignment. Students report times of 25–30 hours.

1. Add a symbol table to your program. Define its structure and a compiler switch to allow the printing of the symbol table after compilation. Note that you will be adding functions and procedures in Part VIII, so your symbol table structure should be extensible.

2. Enhance your program to scan and parse programs described by the grammar of the previous assignment with the addition of declarations and array features as described by the following BNF. Add appropriate declarations to the programs from previous parts and run your compiler on them.

3. Write your routines which access the symbol table as *abstract functions* and *procedures*, that is, the calling procedures, e.g.,

   enter (SymbolTable, Name)

   needn't know what the symbol table data structure looks like.

4. Your symbol table output should:
   • Be **in alphabetical order** by name.
   • Include, at a minimum (more is better), the characters in the name, the class, scoping information (for future enhancement), the line number where defined, line numbers where referenced and any other information which seems relevant.

   For example, if the program contains the following

```
. . .
List : Array [1..100] of integer;
a :     Integer;
. . .
BEGIN
. . . .
END
```

   then the symbol table might be:

Name	**Class**	**Scope**	**Definition Line No.**	**Reference Line Nos.**	**Other**
...					
a	Integer	0	#		
List	Array of Integer	0	#	#, #, #	One Dimensional
...					

NEW BNF (Merge with Old BNF)

Program	$\rightarrow$	DeclarativePart
		begin
		SequenceOfStatements
		end;
DeclarativePart	$\rightarrow$	{BasicDeclarativeItem}
BasicDeclarativeItem	$\rightarrow$	BasicDeclaration
BasicDeclaration	$\rightarrow$	ObjectDeclaration
ObjectDeclaration	$\rightarrow$	IdentifierList : SubtypeIndication;
SubtypeIndication	$\rightarrow$	TypeDefinition | TypeMark
TypeDefinition	$\rightarrow$	ArrayTypeDefinition
ArrayTypeDefinition	$\rightarrow$	ConstrainedArrayDefinition
ConstrainedArrayDefinition	$\rightarrow$	**array [**ConstrainedIndexList**] of** ElementType
ConstrainedIndexList	$\rightarrow$	DiscreteRange { , DiscreteRange}
DiscreteRange	$\rightarrow$	Range
Range	$\rightarrow$	SimpleExpression .. SimpleExpression
ElementType	$\rightarrow$	TypeMark
TypeMark	$\rightarrow$	**integer** | **boolean**
IdentifierList	$\rightarrow$	Identifier { , Identifier}
name	$\rightarrow$	SimpleName | SimpleName {NameSuffix}
NameSuffix	$\rightarrow$	**(**Expression { , Expression}**)**

9

Introduction to Code Generation

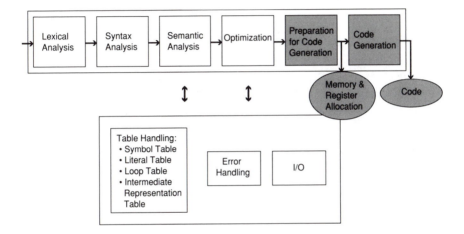

9.0 Introduction

The code generation phase translates the intermediate representation into "code". In this text, the final code will be assembly language code which is then assembled, linked and executed.

The preparation for code generation phase decides how to allocate registers and memory. The code generation phase translates the intermediate representation to assembly language using these registers and memory assignments.

9.1 Preparation for Code Generation

To prepare for code generation, the compiler decides where values of variables and expressions will reside during execution. The preferred location is a register since instructions execute faster when the data referred to in operands reside in registers. Ultimate storage is often a memory location, and due to the scarcity of registers, even intermediate results may need to be assigned memory locations also.

Various methods have been developed for good use of registers. One technique is to store all loop variables in registers (until there are no more registers) since statements inside loops may execute more than once.

141

Another easily implemented technique is to store the variables used the most in registers.

For machines with a stack and a stack pointer, operations involving the stack are generally quicker than those involving an access to (non-stack) memory. Thus, another code generation technique is to push all variables in a procedure onto the stack before the procedure is executed, access them from the stack for code generation and then to pop them off at the end. Example 8 in Chapter 1 used a stack (and two registers).

Careful assignment of expressions and variables to registers can increase the efficiency of the resulting compiler. In this chapter, we will generate code as though there were only one available register. (See the Exercises and Lemone (1992), Chapter 10 for more efficient strategies.)

9.2   Generation of Directives from the Symbol Table

The symbol table is used to generate directives. Thus if *A* is entered in the symbol table with class equal to *variable*, and attribute type equal to *integer*, then the directive which allocates space for an integer is generated.

Example 1 shows directives for integers for a number of machines.

EXAMPLE 1   Directives for a symbol table entry with class = variable, attribute type = integer and character string = *A*

```
On the Vax:          A:      .LONG
On the M68000:       A       DS.L
On the 86-family:    A       DW        ?
```

The VAX-11 and the M68000 directives allocate 32 bits in memory for the variable *A*.

The 86-family (which includes the IBM-PC) directive allocates 16 bits in memory for the variable *A*. None of these directives assigns an initial value (although it is possible to do so). If the symbol table for *A* mentioned that this were a double-sized integer, then other directives would be generated, e.g., DD (double word) on the 86-family.

9.3   Code Generation from Abstract Syntax Trees

A simple code generator can generate code from an abstract syntax tree merely by walking the tree.

Consider the following abstract syntax tree for the two assignment statements:

```
X1 := A + BB * 12;
X2 := A/2 + BB * 12;
```

from Chapter 1:

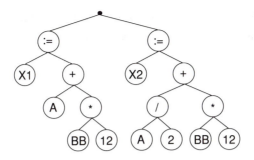

Using a tree walk which generates code for the subtrees first, then applies the parent operator, the following code can be easily produced :

```
Load        BB,R1
Mult        #12,R1
Store       R1,Temp1
Load        A,R1
Add         Temp1,R1
Store       R1,Temp2
Load        Temp2,R1
Store       R1,X1
Load        A,R1
Div         R1,#2
Store       R1,Temp3
Load        BB,R1
Mult        #12,R1
Store       T4,R1
Load        T3,R1
Add         T4,R1
Store       T5,R1
Load        T5,R1
Store       R1,X1
```

For two leaf nodes, the method shown loads the left-most into a register. The right-most leaf node could just as well have been the one put into the register.

Exercise 3 asks the reader to write the algorithm for this method.

Notice that the code continually stores the value in the register into memory in order to reuse it for the next computation. This is called a *register spill*.

The reader may wish to compare this with the code produced in Chapter 1, Section 9. The algorithm there also traversed the tree in the same order. However, more than one register was used and values and addresses for the variables had been pushed onto the stack.

The example in this section generates code for assignment statements whose right-hand sides are expressions. The next section discusses code generation strategies for various other language constructs.

9.4   Standard Code Generation Strategies

A code generator can be written to recognize standard "templates":

(1) Assignment Statements

A tree pattern of the form

generates a Move (copy) instruction:

```
MOVE aPlace,T
```

where *aPlace* represents the register, stack position or memory location assigned to *a*.

(2) Arithmetic Operations

Suppose *Op* represents an arithmetic operation and consider an abstract syntax tree for the statement $t := a \; Op \; b$:

One possible code sequence is:

```
MOVE     aPlace,Reg
OP       bPlace,Reg
MOVE     Reg,T
```

This is the method used in the example of the previous section. Of course, some machines require that special registers be used for some operations such as multiplication and division.

Example 2 shows code for the statement $T := A - B$.

EXAMPLE 2   Code for $T := A - B$

An abstract syntax tree is:

Following the template above yields:

```
MOVE        aPlace,Reg
SUBTRACT    bPlace,Reg
MOVE        Reg,T
```

Here the SUBTRACT instruction subtracts the first operand from the second.

(3) IF Statements

IF statements can be represented by an abstract syntax tree such as:

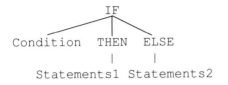

A typical code sequence is

```
          (Code for condition)
          BRANCHIFFALSE                    Label1
          (Code for Statements1)
          BRANCH                           Label2
  Label1: (Code for Statements2)
  Label2:
```

Example 3 illustrates this for the statement:

```
IF a < b THEN Max := b ELSE Max := a
```

EXAMPLE 3   IF a < b THEN   Max := b ELSE Max := a

The abstract syntax tree is:

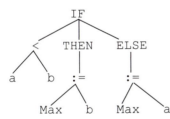

Following the model above, the code would be:

```
        COMPARE    aPlace,bPlace; (Code for comparison)
        BGEQ       Label1        ; NOT <
        MOVE       bPlace,Max    ; (Code for Statements1)
        BRANCH     Label2
Label1: MOVE       aPlace,Max    ; (Code for Statements2)
Label2:
```

In Example 3, *aPlace* and *bPlace* may refer to the variables *a* and *b* themselves, if the machine allows two memory operands. For machines such as the 86-family (which includes the IBM-PC) which require one operand to be in a register, the instruction:

```
COMPARE      aPlace,bPlace
```

can be replaced by:

```
MOVE         aPlace, Reg
COMPARE      Reg,bPlace
```

if neither *a* nor *b* is assigned to a register.

We write all operands as *Source,Destination* although some machines, in particular the 86-family, expect the operands in the order *Destination,Source*.

The instruction:

```
MOVE      bPlace,Max     ; (Code for Statements1)
```

can be replaced by:

```
MOVE      bPlace,Reg     ; (Code for
MOVE      Reg,Max        ; Statements1)
```

Similarly, the instruction at *Label1* can be replaced by two instructions, each using only one memory operand.

(4) Loops

In some sense, loops are just conditionals whose code is repeated. Consider the loop:

```
LOOP While condition DO
      Statements
ENDLOOP
```

An abstract syntax tree is:

```
            LOOP
          /      \
Condition    Statements
```

A reasonable code sequence might be:

```
              (Code for NOT condition)
Label1:  BRANCHIFTRUE            Label2
              (Code for Statements)
              BRANCH                 Label1
Label2:
```

Example 4 shows such a loop.

EXAMPLE 4   Code for WHILE A < B DO X := 3

The abstract syntax tree is:

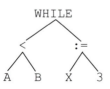

A possible code sequence is:

```
Label1:  COMPARE   APlace , BPlace
         BRANCHGE  Label2
         MOVE      #3,X
         GOTO      Label1
Label2:
```

Here, the NOT has been incorporated into the conditional branch, BRANCHGE, pseudocode for "Branch on Greater Than or Equal To".

9.5 Summary

This chapter is an introduction to code generation and presents a simple tree-walk strategy together with templates for expressions, assignment statements, conditionals and loops.

Code generation is a large case analysis. Each non-leaf node is examined for whether it is an arithmetic operator or some other node type such as assignment or IF, and based on this value, the appropriate instruction is selected.

Improvements in the emitted code can be made by using the stack and by assigning values to registers.

Simple code generation is very easy. Good code generation is very difficult.

EXERCISES

1. Rewrite all the code sequences from Section 9.4 using only one memory reference for each instruction. That is, precede each operation with a *MOVE* of one of the operands into a register and then use that register as one of the operands of the operation.

2. Array References: Most machines today have instructions which perform indexing. Consider the statement:

   ```
   A := B[I]
   ```

   An abstract syntax tree might be:

   (a) Write a code sequence as in Section 9.4 for such array references. (Hint: Consider indirect addressing modes.)
   (b) Do the same for *A[I] := B*.

3. Write the code generation algorithm which is described in Section 9.3.
   (Hint 1: Examine a node (using a case statement) and then make recursive calls to the subnodes.)

(Hint 2: Implement the algorithm for assignment statements first, then add more complex nodes such as those for IF statements.)

4. (a) Create an abstract syntax tree for $((a + b) * (a - b) / ((c + d) * (c - d)))$.
   (b) Generate assembly language code using the algorithm in Exercise 3.

5. Boolean Expressions: The code for conditionals in IF statements and WHILE loops did not store the value of the conditional. Most languages allow conditionals, called Boolean expressions or logical expressions, to appear as the right-hand side of assignment statements. Thus,

```
Flag := a < b
```

is a legal statement if *Flag* has been assigned a type which allows true or false values. The code template for this might store one value in a register if the expression is true and another value if the expression is false:

```
        (Code for boolean)
        BRANCHIFFALSE      Label1
        MOVE               #-1,Reg
        BRANCH             Label2
Label1: CLEAR              Reg
Label2:
```

Using this template and an appropriate abstract syntax tree for the above assignment statement, write assembly language code.

6. More Boolean Expressions:
   (a) Write a short Boolean grammar which allows NOT, AND, and OR as well as the boolean relations $<, >, =, \neq, \leq, \geq$, and parenthesized Boolean expressions.
   (b) Design abstract syntax tree nodes for Boolean expressions.
   (c) Expand the code template from Exercise 4 to handle Boolean expressions
   (d) Use your code templates and the usual tree algorithm to output code for:

```
Flag := a < b OR a < c
```

7. CASE Statements:
   (a) Design code templates for CASE statements such as:

```
            CASE Tag    is
Value 1 :  (Statements 1)
Value 2 :  (Statements 2)
    ...
Value n :  (Statements n)
```

   (b) Use the templates from (a) and the standard code generation algorithm from Exercise 3 to emit code for:

```
     CASE X   is
3 :  X := X + 1;
4 :  X := X - 1;
5 :  X := 0;
```

8. REPEAT Loops: Section 9.4 illustrated code for loops with a WHILE loop. REPEAT loops differ from WHILE loops in that they always execute the code within them at least once.

(a) Design a code template for the following:

```
REPEAT
   Statements
UNTIL Condition
```

(b) Use the template designed in (a) to generate code for:

```
REPEAT
   Sum := Sum + X;
   X := X + 1
UNTIL Sum ≥ 100
```

9. Code Generation from Other IR's: Chapter 7 discusses other intermediate representations such as reverse Polish and quadruples. Using these forms, develop code templates as in Section 9.4.

Compiler Project Part VIII
Code Generation

1. Write a simple code generator for the Ada subset you have implemented.

    (a) You should also do some *simple* register allocation, that is, assign some of the program's variables or expressions to registers. Be sure to describe your method clearly.

    (b) Translate from the intermediate representation (the AST); that is, the abstract syntax tree translation and the symbol table are the input to your program.

    (c) You may translate into any assembly language, but if you know the assembly language for the machine you are using, that is preferable.

    Hint 1: Start by generating code for assignment statements, adding other constructs one at a time.

    Hint 2: A simple *working* code generator is better than an elaborate *non-working* one!!!

2. Run your compiler on various programs. Output should be the assembly language code.

3. If you are on the machine which uses your assembly language, assemble and run (perhaps using the debugger), producing something which shows your code executing correctly. This is the final assignment. Package and document it well!

4. Include documentation describing your language. Be sure to include the BNF for the subset of Ada which your program compiles.

    (Very Difficult) EXTRA CREDIT: Describe the language (SubAda) using an attribute grammar.

Alternate Assignment

Instead of writing a program which translates abstract syntax trees to assembly language, write a program which *interprets* the abstract syntax tree directly. You may wish to implement a simple *Print* statement in order to view results. Alternatively, the values computed by the program for the variables in the program may be stored in a new field in the symbol table. Printing out the symbol table at the end of the program "execution" will show how the program computed.

Extended Assignment

Implement subprograms in Ada using the following BNF which includes productions for procedure and function declarations and their invocation. Note that there is a new Start symbol in the grammar-*CompilationUnit*.

Appendix

Answers to Selected Exercises

Chapter 1

1. (a) – (v)
   (b) – (i)
   (c) – (iv)
   (d) – (iii)
   (e) – (vi)
   (f) – (ii)

2.

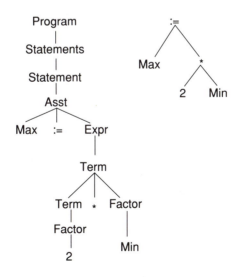

6. Scanner: (Id,"Max") (Op,":=") (Id,"Min") (Op,"+") (Lit,4)
   (Op,*) (Lit,3)

Parser:

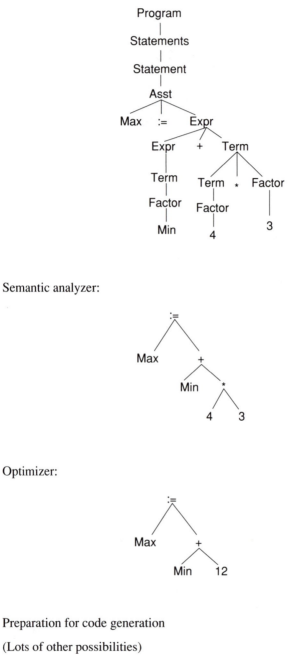

Semantic analyzer:

Optimizer:

Preparation for code generation

(Lots of other possibilities)

```
StackTop  →    |Min              |
               |Address of Max   |
Reg 1 :   Min
```

Code Generation

```
PushAddr   Max
Push       Min
Load       (StackTop),Reg1        Reg1 = Min
Add        #12,Reg1               Reg1 = 12*Min
Store      Reg1,@1(StackTop)      Max = Reg1
```

7. A := 12 * (Ex + El/Em)

8. (a)

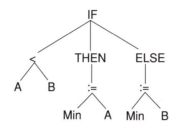

(b)

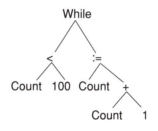

9. (a) Token = Basic lexical unit
   (b) Syntax analysis = Parsing
   (c) Parse tree = Syntax tree
   (d) Intermediate representation = Intermediate code
   (e) Abstract syntax tree = Abstract structure tree
   (f) Analysis phase = Front end

10. (a) Run the X_0 program (that takes X programs to M programs) through the X_0 compiler (the M program which takes X_0 programs to M programs). The output is M code, but it is M code which takes X programs to M programs. Thus, we have a program written in M which takes X programs to M programs. This is an X compiler.
   (b) Take the M program which translates L programs to N code and run the L program which translates L programs to N code through it. The resulting N code is a program which takes L programs to N code. This is an L compiler for machine N.

Chapter 2

7. (a) It is an NFA because there is more than one transition from state 2 upon reading an x.

   (b) A string of one or more x's interspersed with y's, each of which must be followed by at least one x.

   The DFA is :

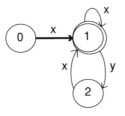

8. (a) Number = Digit$^+$(.Digit$^+$)$^?$ where ? means 0 or 1

   (b)

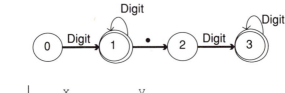

9.

	x	y
1	{1,2,3}	—
2	—	{1}
3	—	—

11. (a) a $(a \mid b)^*$ or a $(a \mid b)^+$ (depending on how "strings of a's and b's" is interpreted)

   (b) $(b^*(a\,b^*\,a)^*\,b^*)^*$

   (c) Palindromes cannot be represented by a regular expression

   (d) $(a\,a \mid a\,b \mid b\,a \mid b\,b)(a \mid b)^*(a\,b)$

12. {aaa}

Chapter 3

1. (a)

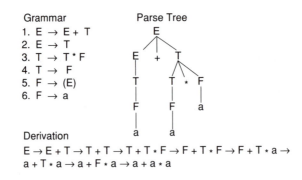

Grammar
1. E → E + T
2. E → T
3. T → T * F
4. T → F
5. F → (E)
6. F → a

Parse Tree

Derivation
E → E + T → T + T → T + T * F → F + T * F → F + T * a →
a + T * a → a + F * a → a + a * a

(b)

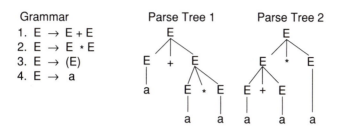

```
Grammar
1. E → E + E
2. E → E * E
3. E → (E)
4. E → a
```

(c)

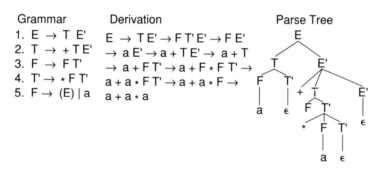

```
Grammar                Derivation
1. E → T {+ T}         E → T + T → F + T
2. T → F {* F}           → a + T → a + F * F
3. F → (E) | a           → a + a * F → a + a * a
```

(d)

```
Grammar                Derivation
1. E → T E'            E → T E' → F T' E' → F E'
2. T → + T E'            → a E' → a + T E' → a + T
3. F → F T'              → a + F T' → a + F * F T' →
4. T' → * F T'           a + a * F T' → a + a * F →
5. F → (E) | a           a + a * a
```

4. An interpret is: time as an imperative verb, to time (clock) flies the same way arrows are timed (clocked).

5. (b) D is the only possible handle because it is the left-most (and longest) string which is the right-hand side of a production and such that all symbols to its right are terminals.
   (c) 5 is the handle since N5 and N are not. The only other choice here is the 5.
   (d) N5 is the handle since it is the longest string which exists on the right-hand side of a production.

7. The string "1 2 3" has more than one parse tree:

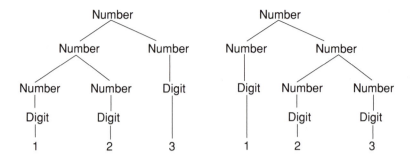

An unambiguous grammar which generates the same set of strings is:

```
Number → Digit | Digit Number
Digit  → 0 | 1 | 2 | ... | 9
```

8. (a) Strings of balanced parentheses (including none)
   (b) Not ambiguous
   (c) Not left recursive

Chapter 4

2. S → C C
   C → a C
   C → b

Stack (Top on left)	Input	Production
S$	a b a b $	S → C C
C C $	a b a b $	C → a C
a C C $	a b a b $	Match
C C $	b a b $	C → b
b C $	b a b $	Match
C $	a b $	C → a C
a C $	a b $	Match
C $	b $	C → b
b $	b $	Match
$	$	Accept

3.

		w	x	y	z
Z → S	Z	Z → S	Z → S	Z → S	
S → w S					
S → A B	S	S → w S	S → A B	S → A B	
A → x A					
A → y	A		A → x A	A → y	
B → z	B				B → z

Stack (Top on right)	Input	Production
$ Z	wwxyz$	Z → S
$ S	wwxyz$	S → w S
$ S w	wwxyz$	Match
$ S	wxyz$	S → w S
$ S w	wxyz$	Match
$ S	xyz$	S → A B
$ B A	xyz$	A → x A
$ BAx	xyz$	Match
$ BA	yz$	A → y
$ By	yz$	Match
$ B	z$	B → z
$ z	z$	Match
$	$	Accept

```
      Z
      |
      S
     / \
    w   S
       / \
      w   S
         / \
        A   B
       / \  |
      x   A z
          |
          y
```

4. S → x
   S → (S R
   R → , S R
   R →)

(a) LL(1) : For the 2 productions S → **x**, S → (S
FIRST (**x**) ∩ FIRST ((S R) = ø
For the 2 productions, R → , S R, R →),
FIRST (, SR) ∩ FIRST()) = ø

(b) Parse table

	x	(	,	)
S	S → **x**	S → (S R		
R			R → , S R	R →)

(c)

Stack(Top on right)	Input	Production
$ S	(x , (x , x)) $	S → (S R
$ R S (	(x , (x , x)) $	Match
$ R S	x , (x , x)) $	S → **x**
$ R **x**	x , (x , x)) $	Match
$ R	, (x , x)) $	R → , S R
$ R S ,	, (x , x)) $	Match
$ R S	(x , x)) $	S → (S R
$ R R S (	(x , x)) $	Match
$ R R S	x , x)) $	S → **x**
$ R R **x**	x , x)) $	Match
$ R R	, x)) $	R → , S R
$ R R S ,	, x)) $	Match
$ R R S	x)) $	S → **x**
$ R R **x**	x)) $	Match
$ R R	)) $	R →)
$ R)	) $	Match
$ R	) $	R →)
$)	) $	Match
$	$	Accept

5. For the grammar

   S → **x** A
   A → **y**
   S → **x** C
   C → **z**

   (a) FIRST(S) = { **x** }

   (b) Not LL(1) since for S → x A, S → x C, FIRST (x A) ∩ FIRST (x C) = {**x**} ≠ ∅

7. (a) The expression grammar

      E → E + T | T
      T → T * F | F
      F → (E) | Id

   is not LL(1) since, for the two productions, E → E + T, E → T,

      FIRST (E + T) = {**(** , **Id**}, FIRST(T) = {**(** , **Id**}
      and {**(** , **Id**} ∩ {**(** , **Id**} ≠ ∅

   (b) The grammar

      E → TE'
      E' → ∈| + T E'
      T → F T'
      T' → ∈| * F T'
      F → (E) | **Id**

   For E' → ∈| + T E', FOLLOW(E') = {)} ∩ FIRST(+ T E') = ∅
   For T' → ∈| * F T', FOLLOW(T') = {+} ∩ FIRST(* F T') = ∅
   For F → (E)| **Id**, FIRST((E)) ∩ FIRST(**Id**) = ∅

   Therefore, the grammar **is** LL(1).

7. FIRST(Program)   = {**begin**}
   FOLLOW(Program)   = { }

   FIRST(SequenceOfStatements)   = {**Identifier**}
   FOLLOW(SequenceOfStatements) =   {**End**}

   FIRST(Statement)   = {**Identifier**}
   FOLLOW(SequenceOfStatements) =   {**Identifier, End**}

   FIRST(SimpleStatement)   = {**Identifier**}
   FOLLOW(SimpleStatement)   = {**Identifier, End**}

   FIRST(Assignment)   = {**Identifier**}
   FOLLOW(SequenceOfStatements) =   {**Identifier, End**}

   FIRST(Name) =   {**Identifier**}
   FOLLOW(Name)   = {:=, +, −, *, /, **mod, rem**,), ; }

FIRST(SimpleName)	=	{**Identifier**}
FOLLOW(SimpleName)	=	{:=, +, −, *, /, **mod, rem**,), ;}

FIRST(SimpleExpression)	=	{**Identifier, Integer,** (}
FOLLOW(SimpleExpression)	=	{), ; }

FIRST(Term)	=	{ **Identifier, Integer,** (}
FOLLOW(Term)	=	{+, −, *, /,), ;}

FIRST(Factor)	=	{ **Identifier, Integer,** (}
FOLLOW(Factor)	=	{+, −, *, /, **mod, rem**,), ;}

FIRST(Primary)	=	{ **Identifier, Integer,** (}
FOLLOW(Primary)	=	{+, −, *, /, **mod, rem**,), ;}

FIRST (AddOp)	=	{ +, − }
FOLLOW (AddOp)	=	{*Name, Integer,* (}

FIRST(MultOp)	=	{*, /, **mod, rem**}
FOLLOW(MultOp)	=	{*Name, Integer,* (}

FIRST(NumericLiteral)	=	{ **Integer,** }
FOLLOW(NumericLiteral)	=	{+, −, *, /, **mod, rem**,), ;}

FIRST(DecimalLiteral)	=	{ **Integer,** }
FOLLOW(DecimalLiteral)	=	{+, −, *, /, **mod, rem**,), ;}

9. A grammar G is LL(k) if and only if

given $A \rightarrow \alpha$, $A \rightarrow \beta$, $\text{FIRST}_k (a) \cap \text{FIRST}_k(\beta) = \emptyset$

where $\text{FIRST}_k(\omega)$ stands for the first k symbols in ω.

10. The grammar
```
S  → i E t S S'
S  → a
S' → e S
S' → ε
E  → c
```

The grammar is not LL(1) since for S' → e S, S' → ε,

```
FOLLOW(S') = {e, $} ∩ FIRST(eS) ≠ ∅.
```

An LL(1) parsing table is:

	a	c	e	i	t	$
S	S → a			S → i E t S S'		
S'			S' → e S S' → ε			S' → ε
E		E → c				

Stack(top on right)	Input	Production
$ S	i c t i c t a e a $	S → i E t S S'
$ S' S t E i	i c t i c t a e a $	Match
$ S' S t E	c t i c t a e a $	E → c
$ S' S t c	c t i c t a e a $	Match
$ S' S t	t i c t a e a $	Match
$ S' S	i c t a e a $	S → i E t S S'
$ S' S' S t E i	i c t a e a $	Match
$ S' S' S t E	c t a e a $	E → c
$ S' S' S t c	c t a e a $	Match
$ S' S' S t	t a e a $	Match
$ S' S' S	a e a $	S → a
$ S' S' a	a e a $	Match
$ S' S'	a e $	S' → e S (choice 1)
$ S' S e	e a $	Match
$ S' S	a $	S → a
$ S' a	a $	Match
$ S'	$	S' → ε
$	$	Accept

Chapter 5

1. List → **Id** [EList]
   EList → **Id**
   EList → EList **, Id**

   Consider the string **Id [Id,Id]**. The first **Id** is the handle with no lookahead. This results in a reduction to EList **[Id,Id]**. This cannot be reduced; thus, the grammar is not LR(0).

2. Sketch of proof:

   The handle in any sentential form in a right derivation from the grammar can be identified by looking one symbol ahead; for example, in $E + T * (T + F)$, the handle E (rather than $E + T$) can be identified by looking one symbol ahead at the "$*$".

7. The new Start symbol is used to create the item *NewStart* → • *Start* which represents the string before parsing has begun.

8. Id • + Id, Id * Id • + Id, Id * Id •+ Id * Id, etc.

9. **State 0:** E' → • E   T → • T * F   F → • Id
             E → • E + T   T → • F
             E → • T   F → • (E)

   State 1: E' → E •
             E → E • + T

State 2: (From step 3 of State Set Algorithm), with $x = T$
          $E \rightarrow T \bullet$ (from State 0)
          $T \rightarrow T \bullet * F$ (from State 0)
          Step 2 doesn't apply

State 3: (From state 0, step 3 of algorithm) with $x = F$
          $T \rightarrow \bullet F$
          Step 2 doesn't apply

State 4: (From state 0, step 3 of algorithm) with $x = ($
          $F \rightarrow (\bullet E)$
          Applying step 2:
          $E \rightarrow \bullet E + T$ $T \rightarrow \bullet F$
          $E \rightarrow \bullet T$ $F \rightarrow \bullet (E)$
          $T \rightarrow \bullet T * F$ $F \rightarrow \bullet Id$
          etc.

14. 0. $S'' \rightarrow S$ (augmented production)
   1. $S \rightarrow i E t S S'$
   2. $S \rightarrow a$
   3. $S' \rightarrow e S$
   4. $S' \rightarrow \epsilon$
   5. $E \rightarrow b$

LR(0) States

State 0	**State 1**	**State 2**
$S'' \rightarrow \bullet S$	$S'' \rightarrow S \bullet$	$S \rightarrow i \bullet E t S S'$
$S \rightarrow \bullet i E t S S'$	(accept state)	$E \rightarrow \bullet b$
$S \rightarrow \bullet a$		

State 3	**State 4**	**State 5**
$S \rightarrow i E \bullet t S S'$	$S \rightarrow a \bullet$	$E \rightarrow b \bullet$
	(FOLLOW(S) = {e, \$})	(FOLLOW(E) = {t})

State 6	**State 7**	**State 8**
$S \rightarrow i E t \bullet S S'$	$S \rightarrow i E t S \bullet S'$	$S \rightarrow i E t S S' \bullet$
$S \rightarrow \bullet a$	$S' \rightarrow \bullet e S$	(FOLLOW(S) = {e,\$})
$S \rightarrow \bullet i E t S S'$	$S' \rightarrow \bullet$ epsilon	

State 9	**State 10**
$S' \rightarrow e \bullet S$	$S' \rightarrow e S \bullet$
$S \rightarrow \bullet i E t S S'$	(FOLLOW(S')={\$})
$S \rightarrow \bullet a$	

(b) Parsing Table

State	ACTION							GOTO		
	i	t	e	a	b	$		S	E	S'
0	S2			S4				1		
1						Acc.				
2					S5				3	
3		S6								
4			R2			R2				
5		R5								
6	S2			S4				7		
7			S9			R4				8
8			R1			R1				
9	S2			S4				10		
10						R3				

Since there are no shift/reduce conflicts, grammar is SLR(1).

(c) Parsing *IF b THEN IF b THEN a ELSE a*

Stack	Input	Action
0	ibtibtaea$	S2
0i2	btibtaea$	S5
0i2b5	tibtaea$	R5
0i2E3	tibtaea$	S6
0i2E3t6	ibtaea$	S2
0i2E3t6i2	btaea$	S5
0i2E3t6i2b5	taea$	R5
0i2E3t6i2E3	taea$	S6
0i2E3t6i2E3t6	aea$	S4
0i2E3t6i2E3t6a4	ea$	R2
0i2E3t6i2E3t6S7	ea$	S9
0i2E3t6i2E3t6S7e9a4	$	R2
0i2E3t6i2E3t6S7e9S10	$	R3
0i2E3t6i2E3t6S7S'8	$	R1
0i2E3t6S7	$	R4
0i2E3t6S7S'8	$	R1
0S1	$	Accept

16.

Nonterminal	Derivation	Leading	Trailing
E	E → E + T	+	+
E	E → T → F → (E)	(	)
E	E → T → T * F	*	*
E	E → T → F → Id	Id	Id
T	T → T * F	*	*
T	T → F → (E)	(	)
T	T → F → Id	Id	Id
F	F → Id	Id	Id
F	F → (E)	(	)

Therefore, Leading and Trailing Sets are:

Nonterminal	Leading	Trailing
E	{ +, *, (, Id }	{ +, *,), Id }
T	{ *, (, Id }	{ *,), Id }
F	{ (, Id }	{), Id }

18. Precedence table for

$S \rightarrow T \mid S \& T$
$T \rightarrow U \mid T \# U$
$U \rightarrow Id$

Leading(S) = {&, #, Id}
Trailing(S) = {&, #, Id}
Leading(T) = { #, Id}
Trailing(T) = { #, Id}
Leading(U) = { Id}
Trailing(U) = { Id}

$S \rightarrow S \& T$

& <• Leading(T)
& <• #
& <• Id

Trailing(T) •> &
•> &
Id •> &

$T \rightarrow T \# U$
<• Leading(U)
<• Id

Trailing (U) •> #
Id •> #

	$	#	Id	$
&	<•	<•	<•	•>
#	•>	•>	<•	•>
Id	•>	•>		•>
$	<•	<•	<•	

Chapter 6

3. 1. S → X X
   2. X → x X
   3. X → y

State	Action			GOTO	
	x	y	$	S	X
0	S3	S4	Err0	1	2
1	Err1	Err2	Acc		
2	S3	S4	Err3		5
3	S3	S4	Err4		6
4	R3	R3	R3		
5	Err5	Err6	R1		
6	R2	R2	R2		

Err0: (a) No input.
(It is only possible to get to Table[1,x] and Table[1,y] after recovery via Err5 and Err6.)

Err1: (a) Unexpected x. End of input expected.
(b) Delete x.

Err2: (a) Unexpected y. End of input expected.
(b) Delete x.

Err4: (a) Unexpected end of input. x or y expected.
(b) Push x or y onto stack then state 3 or 4.

Err5: (a) Unexpected x. End of input expected.
(b) Delete x. Reduce using production 1.

Err6: (a) Unexpected y. End of input expected.
(b) Delete y. Reduce using production 1.

Chapter 7

1. Translate *a := 2 * (Bee + Cee/Dee)* into
   (a) A parse tree. Consider the following grammar:
       Assign   → Var := Expression
       Expression → Expression ± Term | Term
       Term   → Term * Factor | Term / Factor | Factor
       Factor   → (Expression) | **Id**

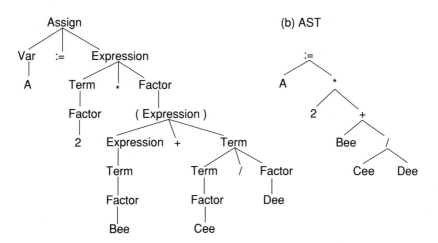

(b) AST

(c) Quadruples

/	Cee	Dee	T1
+	T1	Bee	T2
*	2	T2	T3
:=	T3		A

(d) Triples

(1)	/	Cee	Dee
(2)	+	(1)	Bee
(3)	*	2	(2)
(4)	:=	(3)	A

(e) Indirect triples

(1)	/	Cee	Dee
(2)	+	(1)	Bee
(3)	*	2	(2)
(4)	:=	(3)	A

Execution order: (1) (2) (3) (4)

2. *A[2 * i] := A[i] + 2 * i*

   (a) Abstract syntax tree

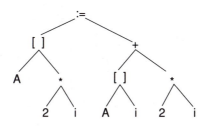

(b) Quadruples

*	2	i	T1
[ ]	A	i	T2
+	T1	T2	T3
*	2	i	T4
[ ]	A	T4	T5
@	T5		T6
:=	T3		T6

(c) Triples

(1)	*	2	i
(2)	[ ]	A	i
(3)	+	(1)	(2)
(4)	*	2	i
(5)	[ ]	A	(4)
(6)	@	(5)	
(7)	:=	(2)	(6)

(d) Indirect triples

(1)	*	2	i
(2)	[ ]	A	i
(3)	+	(1)	(2)
(4)	[ ]	A	(1)
(5)	@	(4)	
(6)	:=	(2)	(5)

Execution order : 1 2 3 1 4 5 6

5. (b)

Syntax	**Semantics**
List → **Id**	List.*NumberOf Dimensions* := 0
List → **Id[EList]**	List.*NumberOf Dimensions* := EList.*NumberOf Dimensions*
EList → E	EList.*NumberOf Dimensions* := 1
$EList_1$ → $Elist_0$, E	$EList_1$.*NumberOf Dimensions* := $EList_0$.*NumberOfDimensions* + 1

(c) *NumberOfDimensions* is a synthesized attribute

(d)

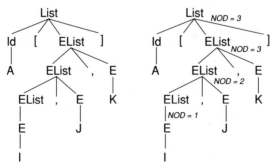

NOD = Number of Dimensions

7. (a) *Use, Def:* Synthesized

   (b) *Live:* Inherited

   (c)

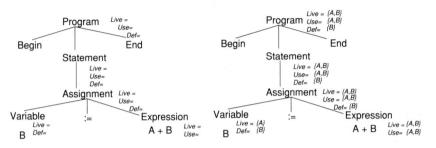

8. $X1 := a + bb * 12$
   $X2 := a/2 + bb * 12$

   (a) **Postfix:**
   X1 a bb 12 * + X1 :=
   X2 a 2 / bb 12 * + :=

   (b) **Quads:**

*	bb	12	T1
+	a	T1	T2
:=	T2		X1

/	a	2	T1
*	bb	12	T2
+	T1	T2	T3
:=	T3		X2

   (c) **Triples**

(1)	*	bb	12
(2)	+	a	(1)
(3)	:=	(2)	X1

(1)	/	a	2
(2)	*	bb	12
(3)	+	(1)	(2)
(4)	:=	(3)	X2

Chapter 8

2. When symbols are in ascending or descending alphabetical order.

4. **type**
```
   tag = record
      next: array [ord("a")..ord("z")] of ^tag;
      size: -2147483648 .. 2147483648;
   vec: ^array [-10..10] of real;
end;
```

var
```
   test : array [1..10] of tag;
```

Chapter 9

2. (a) *A := B[i]:*
```
      LoadAddress   B,Reg
      Add           I,Reg
      Store         (Reg),A
```

(b) *A[i] := B:*
```
      LoadAddress   A,Reg
      Add           i,Reg
      Store         B,  (Reg)
```

3. Procedure CodeGen(Node)
```
   BEGIN
      CASE Node Type is
   Expression Operator, Op : {Result left in Reg1}
      IF neither child is a leaf THEN
      BEGIN
         CodeGen(Left Child);
         Emit "Store, Reg1,T1 := GetTemp";
         CodeGen (Right Child);
         Emit "Store Reg1,T2 := GetTemp"
         Emit "Load Temp1,Reg1"
         Emit "Op Temp2, Reg1"
      END
      ELSE  IF only one child is a leaf THEN
      BEGIN
         CodeGen(Other Child);
         Emit  "Op Leaf Child, Reg1";
      END
      ELSE {Both children are leaves}
      BEGIN
         Emit "Load Left Child, Reg1";
         Emit "Op Right Child, Reg1";
   ":=" : CodeGen (Right Child);
         Emit  "Store Reg1, @Left Child"
```

```
"IF" : CodeGen (First Child);
    Emit "Branch on equal Reg1, Label1 = GetLabel";
    CodeGen (Second Child);
    Emit "Branch Label2 = GetLabel";
    Emit "Label1:";
    CodeGen (Third Child);
    Emit "Label2:"
"WHILE" : Emit "Label1 = GetLabel";
    CodeGen (Left Child);
    Emit "Branch on equal Reg1, Label2 = GetLabel"
    CodeGen(Right Child)
    Emit "Branch Label1"
    Emit "Label2:"
```

4. (b)
```
Load    a, Reg1
Add     b, Reg1
Store   R1, T1
Load    a, R1
Sub     b, R1
Store   R1, T2
Load    T1, R1
Mult    T2, R1
Store   R1, T3
Load    c, R1
Add     d, R1
Store   R1, T4
Load    c, R1
Sub     d, R1
Store   R1, T5
Load    T4, Reg1
Mult    T5, Reg1
Store   Reg1, T6
Load    T3, Reg1
Div     T6, Reg1
```

8. (a)
REPEAT
Statements
UNTIL Condition

```
AST :      REPEAT
          /      \
Statements   Condition
```

Code Template: Label 1: (Code for Statements)
                         (Code for Condition)
                         Branch if Condition false to Label1

BIBLIOGRAPHY

Aho, A., S. C. Johnson, and J. D. Ullman. 1975. Deterministic Parsing of Ambiguous Grammars, *CACM*, 18(8):441–452.

Aho, A. V. 1980. Translator Writing Systems: Where Do They Now Stand?, *IEEE Computer*, August:9–13.

Aho, A. V., R. Sethi, and J. D. Ullman. 1986. *Compilers: Principles, Techniques & Tools*, Reading, MA: Addison-Wesley.
Most instructors, the author included, have grown up (academically) with Aho and Ullman's (now Aho, Sethi and Ullman's) "Dragon Book". It is said that every professional keeps it on his/her shelf.

Aho, A. V., M. Ganapathi, and S. W. K. Tjiang. 1989. Code Generation Using Tree Matching and Dynamic Programming, *ACM TOPLAS*, 11(4):491–516.

Akin, T. A. and R. J. LeBlanc. 1982. The Design and Implementation of a Code Generation Tool, *Software Practice and Experience*, 12:1027–1041.

Allen, F. E. 1970. Control Flow Analysis, *SIGPLAN Notices*, 5(7):1–19.

Babich, W. A. and M. Jazayeri. 1978a. The Methods of Attributes for Data Flow Analysis, Part I. Exhaustive Analysis, *Acta Informatica*, 10:245–264.
Contains a sample attribute grammar for performing data flow analysis on a tree.

Babich, W. A. and M. Jazayeri. 1978b. The Methods of Attributes for Data Flow Analysis, Part II. Demand Analysis, *Acta Informatica*, 10:265–272.
Same as the preceding except the data flow information is computed when needed by the optimizer.

Balachandran, A., D. M. Dhamdhere, and S. Biswas. 1990. Efficient Retargetable Code Generation Using Bottom-Up Tree Pattern Matching, *Computer Languages*, 15(3):127–140.

Barry, R. 1988. An Attribute Grammar for Building Intraprocedural Data Dependence Graphs, Master's thesis, Worcester, MA: Worcester Polytechnic Institute.

Bernstein, D. M., Golumbic, Y. Mansour, R. Pinter, D. Goldin, H. Krawczyk, and I. Nahson. 1989. Spill Code Minimization Techniques for Optimizing Compilers, *SIGPLAN Notices*, 25(7):258–263.

Bickel, M. A. 1987. Automatic Correction to Misspelled Names: A Fourth Generation Language Approach, *CACM*, 30:224–228.

Bochmann, G.V. and P. Ward. 1978. Compiler Writing System for Attribute Grammars, *Computer Journal*, 21(2):144–148.

Briggs, P., K. Cooper, K. Kennedy, and L. Torczon. 1989. Coloring Heuristics for Register Allocation, *SIGPLAN Notices*, 25(7):275–284.

Brown, P. J. 1983. Error Messages: The Neglected Area of Man/Machine Interface, *CACM*, 26(4):246–249.

Burke, M. and G. A. Fisher. 1982. A Practical Method for Syntactic Error Diagnosis and Repair, *SIGPLAN Notices*, 17(6):67–78.

Cattell, R. G. G. 1980. Automatic Derivation of Code Generators from Machine Descriptions, *ACM TOPLAS*, 2(2):173–190.

Chaitin, G. J., et al. 1981. Register Allocation via Coloring, *Computer Languages*, 6:47–57.

Chomsky, N. 1957. *Syntactic Structures*, The Hague: Mouton & Co.

Chow, F. C. 1988. Minimizing Register Usage Penalties at Procedure Calls, *SIGPLAN Notices*, 23(7):85–94.

Cormack, G. V. 1989. An LR Substring Parser for Noncorrecting Syntax Error Recovery, in *Proceedings of SIGPLAN '89 Conference on Programming Language Design and Implementation*, Portland, OR, 161–169.

Courcelle, B. and P. Franchi-Zannettacci. 1982. *Attribute Grammars and Recursive Program Schemes, TCS,* 17(1 and 2):163–191.

Davis, M. D. and E. Weyuker. 1983. Computability, Complexity, and Languages, Fundamentals of Theoretical Computer Science, New York: Academic Press.

Deransart, P., M. Jourdan, and B. Lorho. 1988. *Attribute Grammars: Definitions, Systems, Bibliography*, LNCS 323, Berlin: Springer-Verlag, 83–94.

DeRemer, F. L. and T. Pennello. 1982. Efficient Computation of LALR(1) Look-Ahead Sets, *ACM Trans. on Programming Languages and Systems*, 4(4):615–649.

Dhamdhere, D. M. 1990. A Usually Linear Algorithm for Register Assignment Using Edge Placement of Load and Store Instructions, *Computer Languages*, 15(2).

Donahue, J. Modula-3: A Case Study in Programming Language Design, SIGPLAN '90 Advanced Topics Tutorial.
   Discusses problems in language design, using Modula-3 as an example.

Farrow, R. 1982. Experience with an Attribute Grammar-Based Compiler, in *9th ACM Symp. on Principles of Programming Languages*, 95–107.

Fischer, C. N., J. Mauney, and D. R. Milton, 1980. Efficient LL(1) Error Correction and Recovery Using Only Insertions, *Acta Informatica*, 13(2):141–154.

Fischer, C. N. and R. J. LeBlanc, Jr. 1988. *Crafting a Compiler*, Menlo Park, CA: Benjamin/Cummings.
   This book is on the same level as the "Dragon Book" (see Aho et al., 1986). The chapter on Attribute Grammars is especially well done. There are also excellent sections on error handling and table compaction.

Ganapathi, M. J., J. L. Hennessy, and C. N. Fischer. 1982. Retargetable Compiler Code Generation, *Computing Surveys*, 14(4):573–592.

Ganapathi, M. J. and C. N. Fischer. 1983. Automatic Compiler Code Generation and Reusable Machine-Dependent Optimization—A Revised Bibliography, *ACM SIGPLAN Notices*, 18(4):27–34.

Ganapathi, M. J. and C. N. Fischer. 1984. Attributed Linear Intermediate Representations for Retargetable Code Generators, *Software Practice and Experience*, 14(4):347–364.

Ganapathi, M. and G. O. Mendal. 1989. Issues in Ada Compiler Technology, *IEEE Computer*, 22(2):52–60.
Drawing on experience with designing Ada compilers, the authors describe some implementation issues.

Genillard, C. and A. Strohmeier. 1988. GRAMOL, A Grammar Description Language for Lexical and Syntactic Parsers, *SIGPLAN Notices*, 23(10):103–122.
Using Ada packages for implementation, and an unstratified metalanguage, scanners and either top-down or bottom-up parsers are produced, allowing for a dynamic change of source language.

Ghezzi, C. and M. Jazayeri. 1987. *Programming Language Concepts*, 2nd ed., New York: John Wiley & Sons.

Glanville, R. S. 1977. A Machine Independent Algorithm for Code Generation and Its Use in Retargetable Compilers, Ph.D. thesis, Berkeley: University of California.

Goldberg A. and D. Robson, 1983. *Smalltalk-80: The Language and Its Implementation*, Reading, MA: Addison-Wesley.

Graham, S. L. and R. S. Glanville. 1978. A New Method for Compiler Code Generation, POPL Conference, 231–240.

Graham, S. L., C. B. Haley, and W. N. Joy. 1979a. Practical LR Error Recovery, *SIGPLAN Notices*, 14(8):168–175.

Graham, S. L., W. Joy, and O. Roubine. 1979b. Hashed Symbol Tables for Languages with Explicit Scope Control, *SIGPLAN Notices*, 14(8):50–57.

Graham, S. L. 1980. Table-Driven Code Generation, *IEEE Computer*, August.

Graham, S. L. 1984. Code Generation and Optimization, in *Methods and Tools for Compiler Construction*, B. Lorho, Ed., London: Cambridge University Press.

Gupta, R., M. L. Soffa, and G. Steele. 1989. Register Allocation via Clique Separators, *SIGPLAN Notices*, 24(7).

Hammond, K. and V. J. Rayward-Smith. 1984. A Survey on Syntactic Error Recovery and Repair, *Computer Languages*, 9(7):51–67.

Hecht, M S. 1977. *Flow Analysis of Computer Programs*, New York: Elsevier, North-Holland.

Holub, A. 1990. *Compiler Design in C*, Englewood Cliffs, NJ: Prentice-Hall.
A large (heavy!) thorough book which contains lots of program stubs for compiler pieces. Contains a complete description, including a user manual and the C code for a top-down parser generator and a bottom-up parser generator.

Hopcroft, J. E. and J. D. Ullman. 1979. *Introduction to Automata Theory, Languages and Computation*, Reading MA: Addison-Wesley.

Horspool, R. N. and A. Scheunemann. 1985. Automating the Selection of Code Templates, *Software Practice and Experience*, 15(5):503–514.

Johnson, S. C. 1975. Yacc, Yet Another Compiler Compiler, C. S. Technical Report #32, Murray Hill, NJ: Bell Telephone Laboratories.
Probably the best-known, most often used compiler-compiler.

Knuth, D. E. 1968. Semantics of Context-Free Languages, *Mathematical Systems Theory*, 2(2):127–145.
The seminal article on attribute grammars.

Knuth, D. E. 1971a. Semantics of Context-Free Languages: Correction, *Mathematical Systems Theory*, 5(1):95.
Provides a correction to the proof that all inherited attributes can be mapped to synthesized attributes.

Knuth, D. E. 1971b. Top-down Syntax Analysis, *Acta Informatica*, 1(2):79–110.

Knuth, D. E. 1971c. An Empirical Study of FORTRAN Programs, *Software Practice and Experience,* 1:105–133.

Lemone, K. A. 1985. *Assembly Language and System Programming for the IBM-PC and Compatibles*, Boston: Little-Brown.

Lemone, K. A. 1987. A Language Processing Laboratory, in Proceedings of the 1987 ACM/-CSE Conference, St. Louis.

Lemone, K. A. and M. E. Kaliski. 1987. *Assembly Language Programming for the VAX-11*, Boston: Little-Brown.

Lemone, K. A. 1990. Document Formatting Using Attribute Grammars, Worcester, MA: WPI Technical Report Series, No. 2.

Lemone K. A., J. J. McConnell, M. A. O'Connor, and J. Wisnewski. 1991. Implementing Semantics of Object Oriented Languages Using Attribute Grammars, in Proceedings of the ACM 19th Computer Science Conference, San Antonio, TX.

Lemone, K. A. 1992. *Design of Compilers*, Boca Raton, FL: CRC Press.

Lesk, M. E. and E. Schmidt. 1975. Lex, A Lexical Analyzer Generator, in *UNIX Programmer's Manual 2*, Murray Hill, NJ: AT&T Bell Laboratories.
The Scanner Generator which is often used with YACC (see Johnson, 1975).

Lorho, B., Ed. 1984. *Methods and Tools for Compiler Construction*, London: Cambridge University Press.
A collection of articles covering the compilation process with some emphasis on attribute grammars. The authors are well known for the area about which they write.

Mauney, J. and C. N. Fischer. 1988. Determining the Extent of Lookahead in Syntactic Error Repair, in *ACM Transactions on Programming Languages and Systems*, 10(3):456–469.
Describes a property of a group of look-ahead symbols which results in a reasonable error-repair algorithm.

Meek, B. 1990. The Static Semantics File, *SIGPLAN Notices*, 25(4):33–42.
If you think you know what static semantics is, you may find your "knowledge" challenged in this series of EMail messages.

Mili, A. 1985. Towards a Theory of Forward Error Recovery, *IEEE Transactions on Software Engineering,* SE-11(8):735–748.

Muchnick, S. and N. Jones, Eds. 1979. *Program Flow Analysis: Theory and Applications,* Englewood Cliffs, NJ: Prentice-Hall, 79–101.

Noonan, R. E. *An Algorithm for Generating Abstract Syntax Trees,* 10(3/4):225–236.

O'Connor, M. and K. A. Lemone. 1987. A Method to Improve Testing and Debugging of Robotic Programs in the Language Development Environment, in *Proceedings of the ACM Computer Science Conference,* St. Louis.
Uses attribute grammars to perform incremental editing of robotic programs.

Pagan, F. G. 1981. *Formal Specification of Programming Languages: A Panoramic Primer,* Englewood Cliffs, NJ: Prentice-Hall.

Pai, A. B. and R. Kieburtz. 1980. Global Context Recovery: A New Strategy for Syntactic Error Recovery by Table-Driven Parsers, *ACM Trans. on Programming Languages and Systems,* 2(1):18–41.
This method uses fiducial symbols, not to do panic-mode recovery, but as the beginning of a set of legal tokens to be parsed; used in the context of LL(1) parser generators.

Park, J. C. H., K. M. Choe, and C. H. Chang. 1985. A New Analysis of LALR Formalisms, *TOPLAS,* 7(1):159–175.

Pyster, A. 1980. *Compiler Design and Construction,* New York: Van Nostrand Reinhold.

Pyster, A. 1988. *Compiler Design and Construction in C,* New York: Van Nostrand Reinhold
Emphasizes front-end of compilers (parsing) and does so clearly. Lots of program description for the student's project.

Richter, H. 1985. Noncorrecting Syntax Error Recovery, *ACM Trans. on Programming Languages and Systems,* 7(3):478–489.
Describes a recovery technique—so that compiler can continue finding errors—without attempting a correction to the error.

Roberts, G. H. 1988. Recursive Ascent: An LR Analog to Recursive Descent, *SIGPLAN Notices,* 23(8):23–29.
Defines a procedure called recursive ascent which draws upon the closure operation in Item set creation.

Roberts, G. H. 1990. From Recursive Ascent to Recursive Descent: Via Compiler Optimizations, *SIGPLAN Notices,* 25(4).
A brief paper describing steps for conversion of LR parsers to LL parsers for languages that are basically LL.

Rosenkrantz, D. J. and H. B. Hunt. 1987. Efficient Algorithms for Construction and Compactification of Parsing Grammars, *ACM Trans. on Programming Languages and Systems,* 9(4):543–566.
Relates the syntax and semantic description of a programming language.

Sacks, R. 1989. Detecting Interprocedural Parallelism During Semantic Analysis, Master's thesis, Worcester, MA: Worcester Polytechnic Institute.
Sequential programs are analyzed to find procedures which may be performed in parallel.

Sager, T. 1986. A Short Proof of a Conjecture of DeRemer and Pennello, *ACM Trans. on Programming Languages and Systems,* 8(2):264–271.
A proof for language theory buffs concerning non-LR(k) grammars.

Siewiorek, D. P., G. C. Bell, and A. Newell. 1982. *Computer Structures: Principles and Examples,* New York: McGraw-Hill.

Snelting, G. 1990. How to Build LR Parsers which Accept Incomplete Input, *SIGPLAN Notices,* 25(4):51–58.

Shows changes to shift-reduce actions which allow a reduction upon seeing a prefix of a sentential form. Useful for language-based editors and perhaps for distributed parsing (although the authors don't mention the latter).

Spector, D. 1988. Efficient Full LR(1) Parser Generation, *SIGPLAN Notices,* 23(12):143–150.

Stating that LR(1) grammars are more intuitive to write than LALR(1) grammars, the author argues further that minimal-state LR(1) parsing tables are not much larger than those for LALR(1).

Terry, P. 1986. *Programming Language Translation: A Practical Approach,* Reading, MA: Addison-Wesley.

Many program listings. Covers the front-end of compilers. Separate chapter on language theory.

Treadway, P. L. 1990. The Use of Sets as an Application Programming Technique, *SIGPLAN Notices,* 25(5):103–116.

Discusses implementation of sets (representation and operations) in Pascal, Ada and C as well as an example (an interrupt handler) application.

Tremblay, S. P. and P. G. Sorenson. 1985. *The Theory and Practice of Compiler Writing,* New York: McGraw-Hill.

A thorough text. Topics concerning languages, grammars and language design are discussed initially.

Waddle, V. E. 1990. Production Trees: A Compact Representation of Parsed Programs, *ACM Trans. on Programming Languages and Systems,* 12(1):61–83.

Introduces an intermediate representation more compact than abstract syntax trees.

Waite, W. and G. Goos. 1984. *Compiler Construction,* Berlin: Springer-Verlag.

Combination of introductory material with state of the art and advanced material. A difficult text.

Watson, D. 1989. *High-Level Languages and Their Compilers,* Reading, MA: Addison-Wesley.

An introduction to language translation.

Wirth, N. 1976. *Algorithms + Data Structures = Programs,* Englewood Cliffs, NJ: Prentice-Hall.

Wulf et al. 1980. An Overview of the Production Quality Compiler-Compiler Project, *IEEE Computer,* 13(8):38–49.

Index